The Shanghai Maths Project

For the English National Curriculum

一课一练

T0187291

Practice Book 6A

Series Editor: Professor Lianghuo Fan

UK Curriculum Consultant: Paul Broadbent

Collins

William Collins' dream of knowledge for all began with the publication of his first book in 1819.

A self-educated mill worker, he not only enriched millions of lives, but also founded a flourishing publishing house. Today, staying true to this spirit, Collins books are packed with inspiration, innovation and practical expertise. They place you at the centre of a world of possibility and give you exactly what you need to explore it.

Collins. Freedom to teach.

Published by Collins
An imprint of HarperCollins*Publishers*
The News Building
1 London Bridge Street
London
SE1 9GF

HarperCollins*Publishers*
Macken House
39/40 Mayor Street Upper
Dublin 1
DO1 C9W8
Ireland

Browse the complete Collins catalogue at
www.collins.co.uk

This book is produced from independently certified FSC™ paper to ensure responsible forest management.

For more information visit:
www.harpercollins.co.uk/green

The Shanghai Maths Project (for the English National Curriculum) is a collaborative effort between HarperCollins, East China Normal University Press Ltd. and Professor Lianghuo Fan and his team. Based on the latest edition of the award-winning series of learning resource books, *One Lesson, One Exercise*, by East China Normal University Press Ltd. in Chinese, the series of Practice Books is published by HarperCollins after adaptation following the English National Curriculum.

Practice Book Year 6A has been translated and developed by Professor Lianghuo Fan with the assistance of Ellen Chen, Ming Ni, Huiping Xu and Dr Jane Hui-Chuan Li, with Paul Broadbent as UK Curriculum Consultant.

Series Editor: Professor Lianghuo Fan

UK Curriculum Consultant: Paul Broadbent

Publishing Manager: Fiona McGlade and Lizzie Catford

In-house Editor: Mike Appleton

In-house Editorial Assistant: August Stevens

Project Manager: Karen Williams

Copy Editors: Catherine Dakin and Tracy Thomas

Proofreaders: Tracy Thomas and Steven Matchett

Cover design: Kevin Robbins and East China Normal University Press Ltd.

Cover artwork: Daniela Geremia

Internal design: 2Hoots Publishing Services Ltd

Typesetting: Ken Vail Graphic Design Ltd

Illustrations: QBS

Production: Sarah Burke

Printed and bound by Ashford Colour Press Ltd

Contents

Chapter 1   Revising and improving

1.1 Using symbols to represent numbers

 Learning objective Solve missing number problems

 Basic questions

1 What number does each ⬤ in the following number sentences represent?

(a) ⬤ + 2.8 = 17.2

⬤ = _____

(b) 10.3 − ⬤ = 4.7

⬤ = _____

(c) ⬤ × 6 = 120

⬤ = _____

(d) 1260 ÷ ⬤ = 9

⬤ = _____

(e) ⬤ × ⬤ = 81

⬤ = _____

(f) ⬤ + ⬤ + ⬤ = 96

⬤ = _____

(g) ⬤ + 1.3 + 7.7 = 14

⬤ = _____

(h) 15.2 − 3.4 − ⬤ = 1.6

⬤ = _____

2 The ▲ in each calculation represents the same digit. Find its value.

(a)
```
    6 ▲
  +   ▲
  ─────
    7 2
```
▲ = _____

(b)
```
    ▲ 8
  −   ▲
  ─────
    ▲ 5
```
▲ = _____

(c)
```
    5 ▲
  − ▲ 8
  ─────
    1 5
```
▲ = _____

(d)
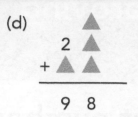

$$\begin{array}{r} 2\,\triangle \\ +\ \triangle\,\triangle \\ \hline 9\ 8 \end{array}$$

$\triangle = \underline{\hspace{2cm}}$

(e)

$$\begin{array}{r} \triangle\ 3 \\ \times\quad \triangle \\ \hline 3\ 7\ 8 \end{array}$$

$\triangle = \underline{\hspace{2cm}}$

(f)
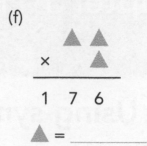

$$\begin{array}{r} \triangle\ \triangle \\ \times\quad \triangle \\ \hline 1\ 7\ 6 \end{array}$$

$\triangle = \underline{\hspace{2cm}}$

(g)

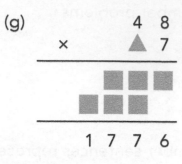

$$\begin{array}{r} 4\ 8 \\ \times\quad \triangle\ 7 \\ \hline \blacksquare\ \blacksquare\ \blacksquare \\ \blacksquare\ \blacksquare\ \blacksquare \\ \hline 1\ 7\ 7\ 6 \end{array}$$

$\triangle = \underline{\hspace{2cm}}$

(h)

$$\begin{array}{r} \triangle\,\overline{)\,2\ 7} \\ 2\ \triangle \\ \hline 2 \end{array}$$

$\triangle = \underline{\hspace{2cm}}$

3 Look for patterns and then write the missing numbers.

(a) 1, 5, 9, 13, ★, 21, 25, 29, ■, 37, ...

★ = ☐ , ■ = ☐

(b) 0.3, 7.7, 0.5, 7.4, 0.7, ●, ★, 6.8, 1.1, 6.5, ...

● = ☐ , ★ = ☐

(c) 2, 4, 8, 16, ■, 64, 128, ★, ...

■ = ☐ , ★ = ☐

(d) 0.1, 0.2, 0.3, 0.5, 0.8, ▲, 2.1, ...

▲ = ☐

4 Complete each statement. (Note: 'natural numbers' are positive integers, that is, 1, 2, 3, ...)

(a) In ⭐ × 71 + ⭐ < 500, the greatest natural number that can be filled in the ⭐ is ____.

(b) In 100 > ■ × 9 > 30, the natural numbers that ■ can stand for are _____.

(c) If 125 ÷ ● = 10 r 5, then the number that ● stands for is ____.

(d) If 12 + ◆ = 3 × ◆, then the number that ◆ stands for is ____.

Challenge and extension questions

5 What digits do the ▲ and ● stand for?

● = ____   ▲ = ____

```
      ▲
      ●
  +  ● ▲
  ───────
    8  9
```

6 Look for patterns and then fill in the ⭕ with suitable numbers.

| 12 | | 35 | 38 | | 25 | 78 | | 49 | 64 | | 71 |

47   63   127   ⭕

11   9   10   ⭕

1.2 Addition and subtraction of decimals (1)

Learning objective Use mental and written methods to add and subtract decimals

Basic questions

1 Work these out mentally. Write the answers.

(a) $0.2 \times 100 =$

(b) $6.3 \div 10 =$

(c) $1.8 + 8.2 =$

(d) $1 - 0.08 =$

(e) $9.6 + 3.04 =$

(f) $25.2 - 5.2 =$

(g) $3.3 + 7.7 =$

(h) $8.8 - 1.5 - 6.5 =$

(i) $9.8 + 0.3 + 9.7 =$

(j) $1 - 0.25 + 0.75 =$

(k) $7.6 \times 10 \div 100 =$

(l) $8.2 - 3.1 + 0.9 =$

(m) $1.2 + 0.18 - 1.2 + 0.18 =$

(n) $0.36 + 0.9 + 0.64 + 8.1 =$

2 Choose the column method to calculate the following. (Check the answer to the question marked with *.)

(a) $98.27 + 2.73 =$

(b) $*2.3 - 0.23 =$

3 Work out the answers to the calculations. Do they give the same answer?

(a) 92.8 − 52.6 + 27.4

(b) 92.8 − (52.6 + 27.4)

4 Calculate smartly.

(a) 5.78 + 4.5 + 4.22

(b) 4.82 + 7.9 − 1.82

(c) 84.67 − (14.67 + 15.3)

(d) 31.2 + 24.58 − 11.2 + 16.42

5 Multiple choice questions. (For each question, choose the correct answer and write the letter in the box.)

(a) When '0' is added to the end of each number, the number that will change value is ☐.

A. 0.24　　　　**B.** 2　　　　**C.** 2.4　　　　**D.** 24.00

(b) When calculating 8.06 ÷ 100 × 10, the result is ☐.

A. 0.806　　　　**B.** 8.06　　　　**C.** 80.6　　　　**D.** 806

(c) In 20.01, if all of the digits are first moved three places to the right across the decimal point and then one place to the left across the decimal point, the result is [ ].

    A. 0.020 01   **B.** 0.2001   **C.** 2.001   **D.** 20.01

(d) Put 5600 m, 5 km 60 m, 5.006 km, 5 km 660 m in order, from the greatest to the least. The second number is [ ].

    A. 5600 m   **B.** 5 km 60 m   **C.** 5.006 km   **D.** 5 km 660 m

6 Convert these units of measurement.

(a) 1.35 kg = [   ] g   (b) 780 kg = [   ] t

(c) 15.4 l = [   ] ml   (d) 30 000 ml = [   ] l

(e) 0.08 m = [   ] cm   (f) 8080 m = [   ] km

(g) 4.2 km² = [   ] m²   (h) 0.5 kg = [   ] g

(i) 1000 cm² = [   ] m²   (j) 150 cm² = [   ] m²

7 Solve these problems.

(a) A rope was originally 10 m long. First, 2.8 m of the rope was cut off and then another 4.2 m was cut off. What is the length of the rope remaining?

[   ]

(b) Tom had £580 in savings. After he spent £60 on stationery and £42.80 on books, how much money did he have left?

[   ]

Challenge and extension question

8 Given $\left(x\right)$ stands for $x + x + x$, and $\boxed{y}$ stands for $y - 0.25$, calculate the following.

(a) $\left(0.95\right)$ + $\boxed{0.8}$

(b) $\left(0.8\right)$ - $\boxed{0.95}$

1.3 Addition and subtraction of decimals (2)

 Learning objective Solve addition and subtraction problems involving decimals

 Basic questions

1 Fill in the missing numbers on the tree diagrams. Write the order in which you completed the calculations below.

(a)

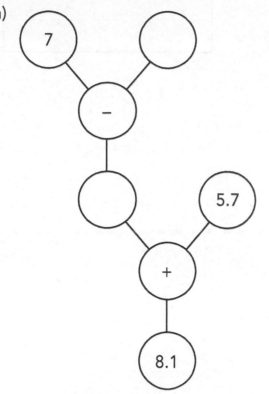

(b)

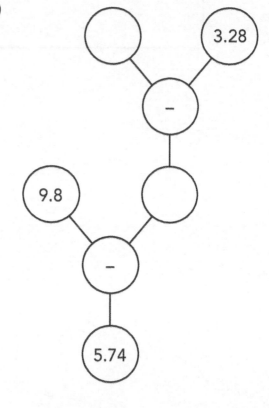

First calculate: _____

_____

then calculate: _____

_____

First calculate: _____

_____

then calculate: _____

_____

2 Draw a tree diagram and then use backward reasoning to find the answer to the following:

Guess the age: Jay and his family were celebrating his grandma's 80th birthday. One guest asked Jay: 'How old are you?' Jay replied with a riddle for the guest to work out: 'If you multiply my age by 6 and then add 8 to it, you will get my grandma's age. How old am I?'

Tree diagram:

Number sentence:

3 Find the missing number.

(a) ☐ − 5.37 + 1.73 = 9

First calculate: _____ then calculate: _____ .

Write the complete number sentence: _____

(b) 19.9 − (☐ + 5.45) = 10.1

First calculate: _____ then calculate: _____ .

Write the complete number sentence: _____

4 Use an efficient method to find out the numbers in each box.

(a) $15.25 + \boxed{} - 0.75 = 18.65$

(b) $29.6 - (\boxed{} - 18.7) = 9.3$

5 Solve these problems.

(a) Some workers were measuring the depth of a river with a 4 m bamboo pole. They put it vertically into the river and found that the part in the mud measured 0.58 m and the part above the water measured 1.27 m. What was the depth of the river?

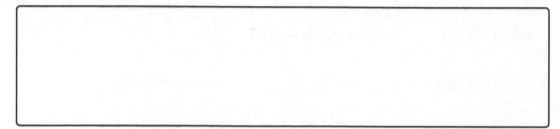

(b) John is 0.03 m taller than Sanjit. Emma is 0.05 m shorter than Sanjit. Emma is 1.42 m tall. How tall is John?

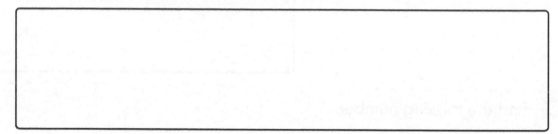

(c) A flour mill produced 85.6 tonnes on the first day. It produced 2.56 fewer tonnes on the second day. On the third day, it produced 6.47 fewer tonnes than on the second day. How many tonnes did it produce on the third day? How many tonnes did it produce in the three days altogether?

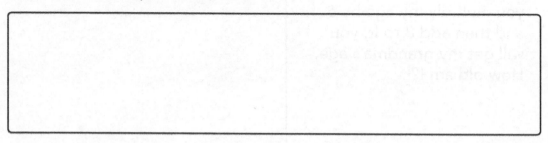

6 There are 48 kg of apples in two baskets. If 4.2 kg of apples are removed from one basket and put into the other basket, the apples in both baskets will have the same weight. How many kilograms of apples does each basket have?

1.4 Decimals and approximate numbers (1)

Learning objective Round decimal numbers to the nearest whole number and to one or two decimal places

Basic questions

1 Fill in the spaces to make each statement correct.

(a) When rounding a decimal number, keeping the whole number means the result is correct to the ones place.

Keeping one decimal place means it is correct to the

_____ place.

Keeping two decimal places means it is correct to the

_____ place, and so on.

(b) When 'rounding off' (or simply 'rounding') a decimal number to a certain place, if the digit in the value place to its right is

_____ than 5, just drop off all the digits to its right.

So, rounding 5.545 to the tenths place, the result is _____.

If the digit is greater than or equal to 5, increase the digit in it by

_____ and drop off all the digits to its right.

So, rounding 10.257 to the hundredths place, the result is

_____.

(c) The '0' in 6.0 should not be dropped when it is rounded to the

_____ place.

2 Round the numbers as required. Fill in the table.

Rounding	1.751	9.995	19.547	23.5023
to the whole number				
to one decimal place				
to two decimal places				

3 The table shows the exchange rates on one day in January 2018. Use this information to answer the questions.

	British Pound (£1)
US Dollar (US $1)	£0.6999
Euro (€1)	£0.7640
Hong Kong Dollar (HK $1)	£0.0898
Chinese Yuan (¥)	£0.1064

(a) €10 = £ [ ]

(b) ¥100 = £ [ ]

4 Based on the exchange rates in the table in Question 3, use rounding to find out the approximate values.

(a) How many £ is 1 HK $ equivalent to? (Round to two decimal places.)

[ ]

(b) How many £ is 100 US $ equivalent to? (Round to one decimal place.)

[ ]

(c) €1 is worth more £ than US $1. How much more? (Round to two decimal places.) [ ]

5 A piece of ribbon was 22.21 m long. 5.9 m was cut off and then the remaining ribbon was cut into 100 equal pieces. How long is each piece? (Keep your answer to the nearest 0.01 m.)

Challenge and extension questions

6 Fill in the missing numbers to make each statement correct.

(a) When 0.999 is rounded to the nearest tenth, it is ⬚ .

The difference between the approximate value and the actual value is ⬚ .

(b) After a decimal number with three decimal places is rounded to the ones place, it is 30. The least possible decimal number is ⬚ .

(c) When a decimal number with two decimal places is rounded to one decimal place, it is 2.7. The greatest possible decimal number is ⬚ . The least possible decimal number is ⬚ .

7 One day in January 2018, Luis went to a bank to exchange €1000 into GBP £ before going shopping. The following were the items, with prices, he wanted to buy.

Headphones: £53.99

Washing machine: £185

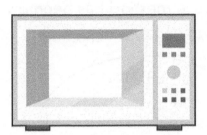

Microwave: £64.99

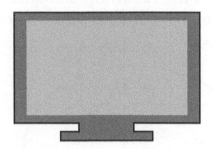

Smart LED TV: £256

Luis is going to buy three items from the above list. How many different ways can Luis buy three items with the money he exchanged? How much money would he have left with each combination? (Refer to Question 3 for the exchange rate.)

1.5 Decimals and approximate numbers (2)

Learning objective Use rounding of decimal numbers to solve problems

Basic questions

1 Round the following amounts of money to whole pence using 'rounding down' or 'rounding up'. One result of each method has been given.

£8.215   £12.2316   £7.998   £99.9124   £35.0080

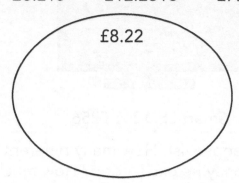

£8.22

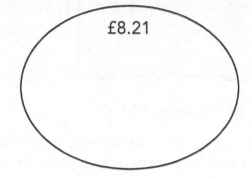

£8.21

Rounding up method   Rounding down method

2 Round the following decimal numbers to one decimal place.

(a)   7.895 ≈ ⬚ (by rounding off)

(b)   33.018 ≈ ⬚ (by rounding up)

(c)   102.087 ≈ ⬚ (by rounding down)

(d)   81.955 ≈ ⬚ (by rounding off)

3 Complete the table.

	11.936	2.4895	1.054
Round down to the whole number			
Round off to one decimal place			
Round up to two decimal places			

4 Multiple choice questions. (For each question, choose the correct answer and write the letter in the box.)

(a) Using the rounding off method, the result of rounding 0.6504 to three decimal places is ☐ .

   A. 0.6504   **B.** 0.65   **C.** 0.650   **D.** 0.651

(b) Using the rounding off method, the result of rounding ☐ to one decimal place is 6.0.

   A. 5.946   **B.** 6.049   **C.** 6.091   **D.** 5.899

(c) When comparing 5.0 and 5, the correct statement is ☐ .

   A. They are the same in value as well as in accuracy.

   B. They are the same in value, but different in accuracy.

   C. They are not the same in value, but the same in accuracy.

   D. They are neither the same in value, nor the same in accuracy.

5 A school bought a red ribbon that was 200 m long. First, 27.6 m was cut off to make 3 pieces of equal length to decorate the hall. The remaining ribbon was cut into 100 equal pieces and given to choir members for their performance. What was the length of each piece of red ribbon given to the choir members? (Round off to 0.01 m.)

6 A rope 52 m long was cut into four pieces. The first piece was 15.2 m long, 3.7 m longer than the second piece. The second piece was 2.8 m shorter than the third piece. What was the length of the fourth piece?

Challenge and extension questions

7 Convert these units of measure and round as indicated.

(a) 2340 kg = [ ] t ≈ [ ] t (Round down to the whole number.)

(b) 31130 ml = [ ] l ≈ [ ] l (Round up to one decimal place.)

(c) 953 m = [ ] km ≈ [ ] km (Round off to two decimal places.)

8 At a book fair, there are 5 different Maths books, 6 different Chinese books and 4 different English books.

(a) Simon wants to buy one of each, a Maths book, a Chinese book and an English book, to make a set. How many combinations can he have?

(b) Simon just wants to buy one resource book from these books. In how many ways can he buy one?

1.6 Revision for circles and angles

Learning objective Draw circles and name and identify parts of circles

Basic questions

1 Use a pair of compasses to draw two circles as indicated, marking the centre O, diameter d and radius r.

(a) radius = 15 mm

(b) diameter = 4 cm

2 The diagram shows a circle with six points.

You can draw [   ] lines by connecting any

two of the six points.

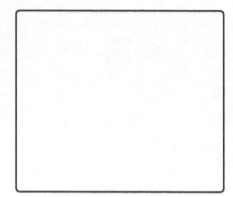

3 The diagram shows a large circle with two

identical small circles inside.

If the diameter of each small circle is

10 cm, then the radius of the large circle is

[   ] cm and its diameter is [   ] cm.

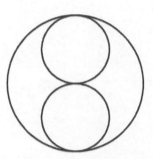

4 Look at the diagram. If $\angle a = \angle b = 40°$, then $\angle c =$ [ ] °.

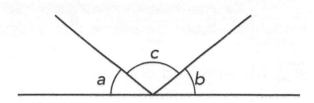

5 In the diagram, $\angle x =$ [ ] °.

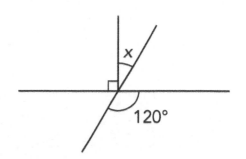

120°

6 Draw the following two figures in the grid below.

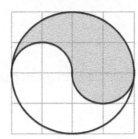 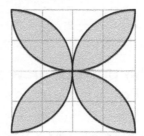

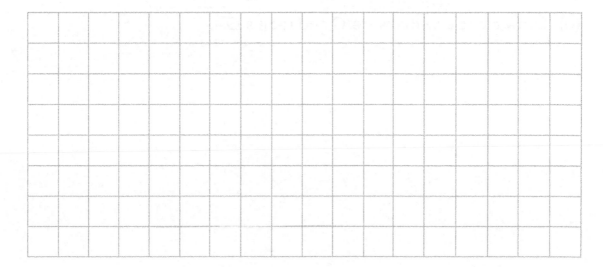

Challenge and extension questions

7 Measure and draw.

(a) Measure the side length of the square shown. It is [   ] cm.

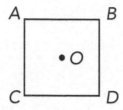

(b) Starting at the point *O*, draw a line *OE* passing through the vertex *C* of the square. (Point *E* is outside the square.)

(c) Measure the size of ∠*ACO*. ∠*ACO* = [   ]°.

(d) Draw a circle with centre *O* and radius *OA*.

Chapter 1 test

1 Work these out mentally. Write the answers.

(a) 0.4 + 0.46 = ⬚

(b) 0.58 + 0.42 = ⬚

(c) 10 ÷ 1000 × 100 = ⬚

(d) 7.5 − 0.25 = ⬚

(e) 5.3 + 4.7 × 10 = ⬚

(f) 5 + 3.04 + 0.96 = ⬚

2 Find the missing number in each calculation. Show your working.

(a) 48.3 − ⬚ + 59.6 = 84.2

(b) 67.3 − (7.68 + ⬚) = 24.4

3 Use the column method to calculate the following. (Check the answer to the question marked with *.)

(a) *35.6 + 149.4 =

(b) 9.07 − 1.88 =

4 Work these out step-by-step. (Calculate smartly, if possible.)

(a) 3.68 + 7.56 − 4.56

(b) 35.6 − 1.8 + 14.4 − 7.2

(c) 30.6 − (10.6 − 5.8) + 4.2

(d) 7.85 + 2.34 + 1.15 + 4.66

(e) 90.8 − 19.28 + 10.72

(f) 0.9 + 0.99 + 0.999 + 0.111

5 Fill in the boxes to make each statement correct.

(a) £1 400 764 ≈ £ ⬚ million (Round to one decimal place.)

(b) £1 947 301 ≈ £ ⬚ million (Round to two decimal places.)

(c) 128 090 678 m ≈ ⬚ million m (Round to the thousandths.)

(d) £1.56 ≈ £ ⬚ (Round to the whole number.)

(e) Using million as the unit, round up the number 90 454 500 to two decimal places. It is ⬚ million.

(f) There are ⬚ decimals with two decimal places that can be rounded off to 8.0. The least of such a number is ⬚ and the greatest of such a number is ⬚ .

6 Multiple choice questions. (For each question, choose the correct answer and write the letter in the box.)

(a) The diagram shows a circle with a radius of 3 cm. The area of the square is ⬚ .

A. 9 cm²   B. 12 cm²

C. 24 cm²   D. 36 cm²

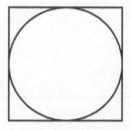

(b) If the diameter of a circle increases by 6 cm, then its radius increases by ☐ cm.

    A. 2   **B.** 3   **C.** 6   **D.** 12

(c) An angle of 210° is ☐ less than a full angle. An angle of 95° is ☐ greater than a right angle.

    A. 30°   **B.** 50°   **C.** 5°   **D.** 150°

(d) When an angle is viewed using a magnifying glass, the angle ☐.

    A. is larger than the original one

    B. is smaller than the original one

    C. is the same as the original one

    D. all the above answers are possible

7 Rewrite the following into decimals with two decimal places.

(a) $7.8 =$ ☐   (b) $4.100 =$ ☐

8 Convert these units of measure.

(a) $2.8 \text{ m} =$ ☐ cm   (b) $138 \text{ ml} =$ ☐ l

(c) $14\,256 \text{ cm}^2 =$ ☐ m^2 (d) $1.52 \text{ km}^2 =$ ☐ m^2

(e) $£4.43 =$ ☐ pence   (f) $7.08 \text{ kg} =$ ☐ kg ☐ g

9 Look for patterns and write what each symbol stands for.

(a) 1, 2, 4, 7, 11, ●, 22, 29, ■, 46, . . .

● = _____   ■ = _____

(b) 19, 9, 17, 8, 15, 7, ▲, ◆, 11, 5, . . .

▲ = _____   ◆ = _____

10 Use a pair of compasses to draw the largest possible circle on the square grid.

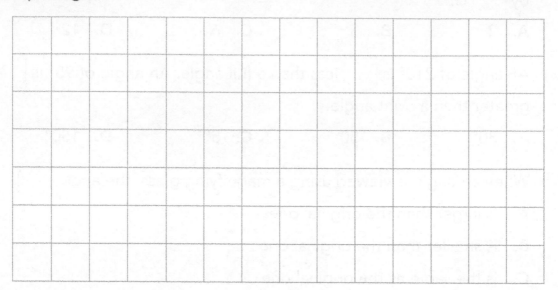

11 Find the unknown angles in the diagrams.

(a) Look at the diagram.
Given ∠BOA = 123°, find ∠BOC.

BOC = ☐ °

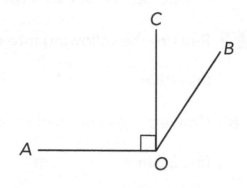

(b) Look at the diagram and find the size of the angle marked with '?'.

? = ☐ °

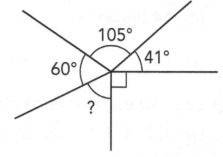

(c) Look at the diagram and find the size of the angle marked with '?'.

? = ☐ °

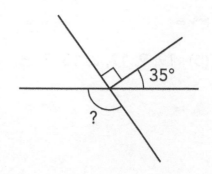

12 The area of the ocean on the surface of the Earth is about 3.61 hundred million km², which is 2.12 hundred million km² greater than the land area. What is the total area of the surface of the Earth?

13 A warehouse had 38 tonnes of food to be delivered to a disaster-hit area. In the first delivery, 7.25 tonnes of food were dispatched. In the second delivery, it was 1.2 tonnes less. In the third delivery, the amount of food dispatched was 1.45 tonnes more than in the second delivery. How many tonnes of food were dispatched in the third delivery? How many tonnes of food were left after the three deliveries?

14 In a long jump game, Andy jumped 2.84 m, which was 0.55 m further than John. Matt jumped 0.17 m further than John. How far did Matt jump?

15 An electrical appliance store had a promotion on a special brand of TV. The original price was £880. The price was first reduced by £80.50, but the sale did not go well. A second price reduction of £102.90 then followed. What was the final price of the TV?

16 Two pieces were cut from a wood fence that was 8.4 m long. The first piece was 2.8 m long, which was 1.9 m shorter than the second piece. How long was the remaining part of the fence?

17 A school was organising an autumn outing, so Jim went to a shop near the school to buy food. The food was priced as follows:

Sesame biscuits: £7.55 per pack	Fruit juice: £4.50 per pack of bottles
Cream biscuits: £8.68 per pack	Cola: £3.60 per pack of bottles
Sandwiches: £2.80 per pack	Sausages: £9.80 per pack

(a) Jim wanted to buy one pack of cream biscuits, one pack of sausages, and one pack of bottles of fruit juice. How much would all the items cost in total?

(b) If Jim wanted to buy one pack of sesame biscuits, one pack of sandwiches, and one pack of bottles of cola, would £15 be enough for him? If it is enough, how much would be left? If it is not enough, how much more does he need?

18 Write the number that each symbol stands for.

(a)

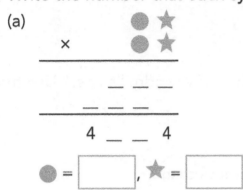

● = [＿＿＿] , ★ = [＿＿＿]

(b) If ▲.■ + ■.▲ = 15.4, then the least value of ▲.■ = [＿＿＿].

(c) If ▲ × ▲ + ▲ ÷ ▲ = 50, ● + ● − ▲ = 2011, then:

▲ = [＿＿＿] , ● = [＿＿＿].

(d) If (■ + ■) + (■ − ■) + (■ × ■) + (■ ÷ ■) = 400,

then ■ = [＿＿＿].

Chapter 2 Multiplication and division of decimals

2.1 Multiplying decimal numbers by whole numbers (1)

Learning objective Multiply whole numbers and decimals with up to two decimal places

Basic questions

1 A toy windmill costs £5.80. How much do 8 toy windmills cost? Use the pupils' strategies below to find the answer.

(a) Jason: 'Let me estimate first.'

8 × ☐ = £ ☐ . So it must be less than £ ☐ .

(b) Tom: 'I do it by converting units.'

£5.8 = ☐ p, and 8 × ☐ = ☐ p.

(c) May: 'I change it to multiplication of two whole numbers.'

8 × 5.8 = ☐

↓ × 10   ↑ ÷ 10

8 × ☐ = ☐

That is: 8 × 5.8

= 8 × 58 ÷ 10

= ☐ ÷ ☐

= ☐

2 Calculate the following.

(a) 8 × 3.2

= 8 × 32 ÷ 10

= [ ] ÷ [ ]

= [ ]

(b) 0.62 × 4

= [ ] × 4 ÷ [ ]

= [ ] ÷ [ ]

= [ ]

(c) 9 × 0.135

= [ ] × [ ] ÷ [ ]

= [ ] ÷ [ ]

= [ ]

3 Estimate first and then calculate the answers.

(a) 8 × 0.94 =

Estimation: _____

Calculation: _____

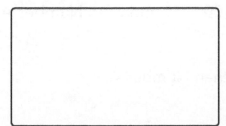

(b) 9.05 × 7 =

Estimation: _____

Calculation: _____

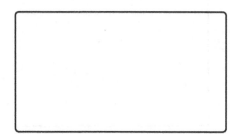

(c) 110.7 × 3 =

Estimation: _____

Calculation: _____

(d) 284.55 × 9 =

Estimation: _____

Calculation: _____

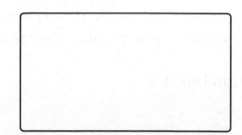

4 Solve these problems.

(a) The side length of a square flowerbed is 3.68 m. If it is fenced on all its sides, how long is the fence in total?

(b) The price of a pencil is £0.75. James bought 9 pencils. How much did he pay?

(c) The ground level of a 6-storey building is 4 m high, and each of the other five levels is 3.6 m high. What is the height of the building?

Challenge and extension question

5 Compare the numbers. Fill in the space with 'greater' or 'less'.

Number A ÷ 100 × 10 = Number B × 100 ÷ 10 and both Number A and Number B are greater than zero.

Number B is _____ than Number A.

2.2 Multiplying decimal numbers by whole numbers (2)

Learning objective Multiply whole numbers and decimals with up to three decimal places

Basic questions

1 Convert the decimal multiplication into whole-number multiplication and then calculate the answer.

```
      4 2                              4 2
  × 0 . 5 7    ┌──────────┐   ×         5 7
              └──────────┘
  ┌──────────┐                    ┌──────────┐
  └──────────┘                    └──────────┘

  ┌──────────┐                    ┌──────────┐
  └──────────┘                    └──────────┘
  ──────────                      ──────────
  ┌──────────┐   ┌──────────┐     ┌──────────┐
  └──────────┘   └──────────┘     └──────────┘
```

2 Use whole-number multiplication first and then calculate the answers to the decimal multiplications.

(a)
```
      2 9
  ×     6
  ────────
  ┌────────┐
  └────────┘
```
(b)
```
      2 9
  × 0 . 0 6
  ────────
  ┌────────┐
  └────────┘
```
(c)
```
      4 7
  ×   1 5
  ────────
  ┌────────┐
  └────────┘
```
(d)
```
  0 . 4 7
  ×   1 5
  ────────
  ┌────────┐
  └────────┘
```

3 Use the fact that 23 × 75 = 1725 to write the products of these multiplications.

(a) 23 × 0.75 = ┌──────┐

(b) 0.023 × 75 = ┌──────┐

(c) 2.3 × 0.075 = ┌──────┐

(d) 2.3 × 75 = ┌──────┐

(e) 2.3 × 750 = ┌──────┐

(f) 75 × 0.23 = ┌──────┐

4 Choose the column method to calculate the following. The first one has been done for you.

(a) 7 × 0.24 = 1.68

```
    0 . 2 4
  ×       7
  ─────────
    1 . 6 8
```

(b) 3.85 × 13 =

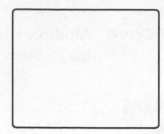

(c) 14.5 × 18 =

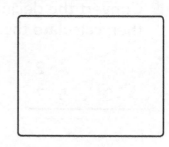

(d) 25 × 0.306 =

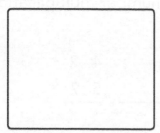

(e) 10.2 × 54 =

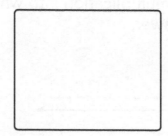

(f) 23.6 × 50 =

5 Are these calculations correct? (Put a ✓ for yes and a ✗ for no in each box and make corrections if necessary.)

(a)
```
        4 . 5
  ×         8
  ───────────
      3 . 6 0
```
☐

(b)
```
      1 . 3 6
  ×   2   5 0
  ───────────
      6   8 0
    2 7 2
  ───────────
    3 4 . 0 0
```
☐

(c)
```
        3 . 1 4
  ×   1 0 5 0
  ─────────────
      1 5 7 0
    3 1 4
  ─────────────
  4 7 1 . 0 0
```
☐

6 Solve these problems.

(a) If an object weighs 1 kg on Earth, it will weigh 0.16 kg on the Moon. Jack weighs 39 kg on Earth. If he was on the Moon, how much would he weigh?

(b) The price of a pen is £12.70 and the price of a notebook is £4.50. Bob wants to buy a pen and two notebooks with £20. Does he have enough money to buy them? Why or why not?

7 Number A is 9 less than Number B. If all of the digits in Number A are moved one place to the right across the decimal point, it becomes 0.009.

Number B is [   ] .

8 Calculate the product of these decimal numbers.

(a) 2.5×0.8 (b) 1.3×3.6 (c) 10.52×0.25

2.3 Addition, subtraction and multiplication with decimals

Learning objective Solve problems with mixed operations, including using brackets

Basic questions

1 Combine two number sentences with a single operation into one number sentence with different operations. The first one has been done for you.

(a) $2.5 + 5.6 = 8.1$

$11 \times 8.1 = 89.1$

(b) $6.2 - 2.6 = 3.6$

$25 \times 3.6 = 90$

(c) $0.45 \times 1.2 = 0.54$

$34 \times 0.54 = 18.36$

$11 \times (2.5 + 5.6) = 89.1$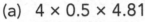

2 Work these out step by step.

(a) $4 \times 0.5 \times 4.81$

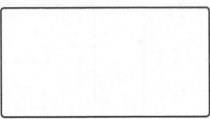

(b) $57.82 - 1.03 \times 42$

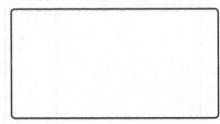

(c) $18 \times (8.14 - 3.64)$

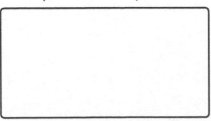

(d) $0.8 \times 50 \times 0.07$

(e) $12.49 - 0.48 \times 25 + 6.3$

(f) $0.75 \times 14 \times 42$

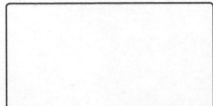

3 Write the number sentences and then calculate the answers.

(a) The sum of 7.8 and 1.2 is multiplied by 0.6. What is the product?

_____

(b) The number that is 1.2 less than 3.2 is multiplied by 3.9. What is the product?

_____

(c) How much more is 10 times 0.44 than 2.5?

_____

(d) Number A is 6.9, which is 0.9 more than twice Number B. What is Number B?

_____

4 Solve these problems.

(a) A textile factory produces clothes for both children and adults. To make one set of children's clothes, it needs 2.08 metres of cloth; to make one set of adult clothes, it needs 3 times as much. To make one set each of children's clothes and adult clothes, how many metres of cloth are needed in total?

(b) A piece of iron rod is 6 m long. It weighs 4.9 kg per metre. How much do 70 pieces of iron rod weigh in kilograms and in tonnes? (Note: 1 tonne = 1000 kg)

(c) Tom bought eight 250 ml bottles of soft drink at £0.32 per bottle. He paid the cashier £4. How much change should he get?

(d) The side length of a square is 0.5 m. Four such squares make one large square. What are the perimeter and the area of the large square?

 Challenge and extension question

5 Insert brackets in the following number sentences to make each equation true.

(a) 15.2 + 2.5 × 3 − 10.6 = 42.5

(b) 30 − 11.8 × 2 − 1.5 = 9.1

(c) 4.5 × 6.5 − 2.5 + 1.8 = 19.8

2.4 Laws of operations with decimal numbers

Learning objective Solve problems with mixed operations, including using brackets

Basic questions

1 Draw lines to match the calculations with the same answers. Use a ruler.

4 × 5.4	8 × 1.25 + 8 × 12.5
4.6 × 9 + 5.4 × 9	6.7 × (4 × 2.5)
8 × (1.25 + 12.5)	(4.6 + 5.4) × 9
6.7 × 4 × 2.5	5.4 × 4

2 True or false? (Put a ✓ for true and a ✗ for false in each box.)

(a) 19 × 0.59 = (19 + 0.1) × 0.59 ☐

(b) 12 × 8.8 = 12 × 8 × 12 × 0.8 ☐

(c) 10 × 2.7 − 5.4 = 2.7 × (10 − 2) ☐

(d) 15 × 2.4 = (15 × 8) × 3 ☐

3 Use the most appropriate method to find the answer to each calculation.

(a) 12.5 × 3.2 × 8   (b) 7 × 0.25   (c) 0.99 × 11

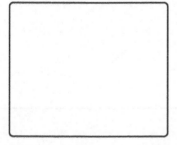

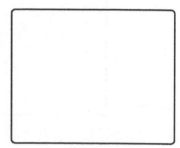

(d) $6.4 \times 6 + 6.4 \times 4 - 6.4$

(e) $720 \times 12.5 \div 8$

(f) $(6.2 \times 5 + 1.8 \times 5) \times 1.25$

(g) $(2.5 + 2.5 + 2.5 + 2.5 + 2.5) \times 88$

(h) $5.4 \times 4 - 7 \times 4.4 + 3 \times 5.4$

(i) $17.48 \times 38 + 1.748 \times 820 - 174.8 \times 2$

4 Solve these problems.

(a) In a school uniform, the blazer costs £12.50 and the trousers cost £10.99. How much do 320 sets of the school uniform cost?

(b) One water pump can pump 12.5 tonnes of water into a swimming pool in one hour. At this rate, how many tonnes of water in total will 16 water pumps pump in 2.5 hours?

Challenge and extension question

5 Write a suitable number in each box to make the equation true.

$4.7 \times 125 + \boxed{} \times \boxed{} = 1000$

2.5 Division of decimals by whole numbers (1)

Learning objective Divide decimals with up to two decimal places by whole numbers

Basic questions

1 Fill in the missing numbers.

(a) $8.4 \div 6 = \boxed{}$

$\boxed{}$ times 0.1 is 8.4.

$84 \div 6 = \boxed{}$

$\boxed{}$ times 0.1 is $\boxed{}$.

(b) $9.38 \div 7 = \boxed{}$

$\boxed{}$ times 0.01 is 9.38.

$938 \div \boxed{} = 134$

$\boxed{}$ times 0.01 is $\boxed{}$.

2 Calculate the following. Most of the first one has been done for you.

(a) $22.4 \div 14 =$

(b) $26.1 \div 9 =$

(c) $41.04 \div 12 =$

22.4 is 224 times 0.1.

```
        1 6
   14 ) 2 2 4
        1 4 0
        ─────
          8 4
          8 4
        ─────
            0
```

16 times 0.1 is $\boxed{}$.

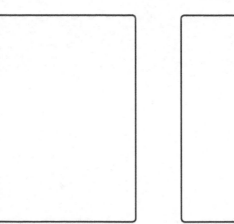

3 Use the column method to calculate the answers. (Check the answers to the questions marked with *.)

(a) 43.2 ÷ 4 =

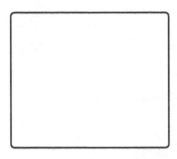

(b) 115.2 ÷ 18 =

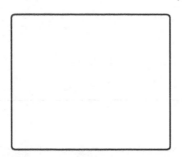

(c) *408.8 ÷ 73 =

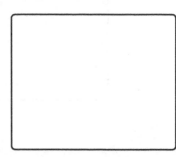

(d) 38.22 ÷ 7 =

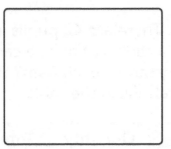

(e) 313.6 ÷ 49 =

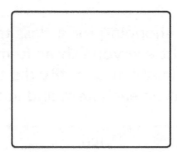

(f) *99.32 ÷ 13 =

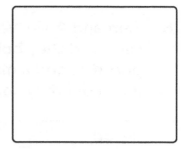

4 Write the number sentences and then calculate the answers.

(a) 6 times a number is 128.4. What is the number?

_____

(b) Number A is 2.28. Number A is 12 times Number B. What is the difference between Number A and Number B?

_____

5 Solve these problems.

(a) A car travelled from place A to place B in 3.6 hours at a speed of 48 km per hour. When the car travelled back from place B to place A, it took 0.4 more hours. What was the speed of the car on the way back?

(b) A piece of wire can be bent into a rectangle of 1.6 m × 1.2 m. If it is bent into a square, what is the area of the square?

Challenge and extension question

6 Sam and Alvin went shopping for a class trip. There are 42 pupils in the class and they bought everyone three items: a carton of milk, a can of oat porridge and a muffin. What quantity did they buy of each item? How much did they spend on each item and in total? Fill in the table.

Food	Price	Quantity	Total price
milk	6 cartons in a box; £0.99 per box		
oat porridge	3 cans in a pack; £4.89 per pack		
muffin	£0.36 each		
Total cost (£)			

2.6 Division of decimals by whole numbers (2)

 Learning objective Solve division problems involving decimal numbers

 Basic questions

1 Calculate mentally and then write the answers.

(a) $6.9 \div 3 =$ ☐

(b) $8.4 \div 4 =$ ☐

(c) $44.8 \div 7 =$ ☐

(d) $59.4 \div 9 =$ ☐

(e) $4.2 \div 6 =$ ☐

(f) $37 \div 5 =$ ☐

(g) $11.8 \div 2 =$ ☐

(h) $8.68 \div 7 =$ ☐

2 Complete the working of the calculation by filling in the blanks.

```
    ☐                    0                  0 . 1 ☐
5 ) 0 . 9 5    →    5 ) 0 . 9 5    →    5 ) 0 . 9 5
                                              5
                                           ─────── ······ represents 45
                                            4  5
                                            4  5      times   ☐
                                           ─────
                                              0
```

The whole number part of the dividend is less than the divisor. The ones place of the quotient should be ☐.

The decimal point in the quotient must be aligned with the decimal point in the _____ .

The method is the same as the division of whole numbers.

3 Use the column method to calculate the answers. (Check the answers to the questions marked with *.)

(a) 6.64 ÷ 8 =

(b) 28.95 ÷ 15 =

(c) *1.248 ÷ 26 =

(d) 8.64 ÷ 36 =

(e) 8.12 ÷ 7 =

(f) *1.11 ÷ 37 =

4 Write the number sentences and then calculate the answers.

(a) 180 times 0.6 is added to the quotient of 62.5 divided by 25. What is the sum?

_____

(b) The difference between 53.8 and 53.26 is divided by 54. What is the quotient?

_____

5 Solve these problems.

(a) An elephant weighs 9.9 tonnes, which is 18 times the weight of a horse. A hippopotamus weighs 3 times as much as a horse. What is the weight of the hippopotamus?

(b) A box of milk has 8 packs and costs £24.80 in total. What is the unit price for one pack of milk? There is a promotion of 'buy 1 box, get 2 packs free' in a supermarket. How much does each pack of milk actually cost? How much cheaper is the promotional price than the original price?

(c) A sugar refinery produced 3.25 tonnes of sugar by using 25 tonnes of sugar cane. How many tonnes of sugar can be produced from every tonne of sugar cane on average? Based on this calculation, how many tonnes of sugar can be produced from 40.5 tonnes of sugar cane?

Challenge and extension question

6 A group of 50 pupils plan to go boating. A 6-seater boat costs £20 while a 4-seater boat costs £15. Propose a few boat renting plans and then complete the following table.

Plan	6-seater boat (No.)	4-seater boat (No.)	Cost (£)
1			
2			
3			
4			

Which renting plan is the cheapest? How much does it cost?

2.7 Division of decimals by whole numbers (3)

 Learning objective Solve division problems involving decimal numbers

 Basic questions

1 A 1.5-litre bottle of juice is shared equally between 5 people. How many litres of juice does each person get?

If it is shared equally between 4 people, how many litres does each get?

1.5 ÷ 5 = ☐

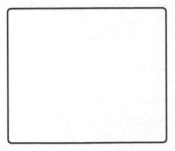

1.5 ÷ 4 = ☐

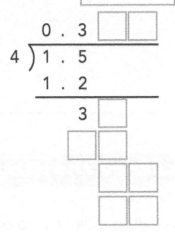

2 Choose the column method to calculate the following. (Check the answers to the questions marked with *.)

(a) 7.8 ÷ 4 =

(b) *26.1 ÷ 6 =

(c) 5.1 ÷ 60 =

(d) 0.9 ÷ 30 =

(e) 5.98 ÷ 52 =

(f) *4.2 ÷ 24 =

3 Write the number sentences and then calculate the answers.

(a) The product of 7.2 and 18 is divided by 4. What is the quotient?

_____

(b) The quotient of 1.04 divided by 26 is subtracted from the sum of eight hundred 0.0125s. What is the difference?

_____

4 Solve these problems.

(a) A clothes shop has made 40 sets of clothes of the same size with 100 metres of cloth. How many metres of cloth were used for one set of clothes?

(b) A 0.24 m long wire is used to form a square. How long is the side of the square? What is the perimeter of the square?

(c) One box of Danish cookies weighs 500 g and costs £2.88, while one box of Continental cookies weighs 800 g and costs £3.88. Which type of cookie is the best value?

Challenge and extension questions

5 A supermarket is selling t-shirts in two different-sized packs: (a) a pack of 5 t-shirts is £19, and (b) a pack of 8 t-shirts is £28. What is the maximum number of t-shirts a customer can buy with £122?

6 During the holidays, many supermarkets offer promotions. Ms Durrani bought four 2-litre bottles of olive oil at £5.45 per bottle. Ms Smith also bought four 2-litre bottles of the same olive oil at £6.12 per bottle, but with a promotion to 'buy three, get one free'. Who had a better deal? How many methods can you find to make the comparison?

2.8 Division of decimals by whole numbers (4)

Learning objective Solve division problems involving decimal numbers

Basic questions

1 Use the fact that **54 ÷ 24 = 2.25** to write the quotients of these division calculations.

(a) 0.54 ÷ 24 = ☐

(b) 5400 ÷ 24 = ☐

(c) 540 ÷ 24 = ☐

(d) 5.4 ÷ 24 = ☐

(e) 0.0054 ÷ 24 = ☐

(f) 0.054 ÷ 24 = ☐

2 Complete the working.

(a)
```
        6
    ┌───────
  4 ) 2  7
     2  4
    ───────
        3  0
```

(b)
```
        0 .
    ┌───────
  8 ) 6 . 0
```

3 Use the column method to calculate the answers.

(a) 36 ÷ 90 =

(b) 8 ÷ 32 =

(c) 10 ÷ 125 =

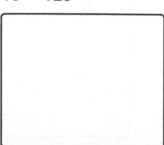

(d) 27 ÷ 72 =

(e) 1 ÷ 16 =

(f) 34 ÷ 8 =

4 Write the number sentences and then calculate the answers.

(a) How many times 26 is 65?

_____

(b) The product of 3 and 0.5 is divided by 12. What is the quotient?

_____

5 Solve these problems.

(a) The weight of an elephant is 4.5 tonnes, which is 90 times as much as an ostrich. What is the weight of an ostrich?

(b) The height of the skyscraper at 30 St Mary Axe in London, also known as the Gherkin, is 180 m. The height of the Shard is 306 m. How many times the height of the Gherkin is that of the Shard?

(c) Ayesha bought 9 exercise books and Mary bought 5 exercise books, each at the same price. Mary spent £31.80 less than Ayesha. How much did each exercise book cost? How much did they spend in total?

6 A store had a promotion for a brand of soft drink, 'buy three 2-litre bottles and get two 650-millilitre bottles free'. Jamal bought six 2-litre bottles of the soft drink. How many millilitres of the soft drink in total should he get?

7 150 pupils from Year 6 went on a school trip to a Science and Technology Centre. The admission ticket was £3 per person plus £0.25 for insurance per person. Three coaches were hired at the cost of £144 each. What was the cost of the visit for each pupil?

2.9 Calculation with calculators

Learning objective Use a calculator to explore patterns and calculate

Basic questions

1 Use the column method to calculate the following, then check the answers with a calculator.

(a) 4.38 × 65 =

(b) 0.978 × 36 =

(c) 96.32 × 16 =

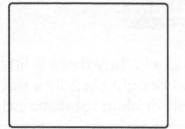
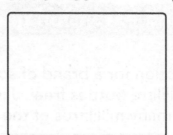
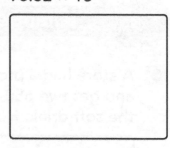

(d) 934.72 ÷ 25 =

(e) 298.15 ÷ 67 =

(f) 443.7 ÷ 15 =

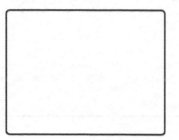
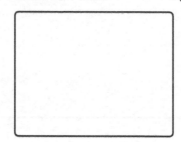
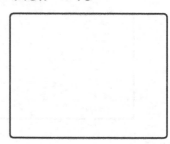

2 Use a calculator to work out questions (b)–(d). Use reasoning to find the answers to questions (e) and (f).

(a) 6 × 6 =

(b) 66 × 66 =

(c) 666 × 666 =

(d) 6666 × 6666 =

(e) 66 666 × 66 666 =

(f) 666 666 × 666 666 =

3 Use a calculator to answer the first four questions below. Explore the pattern and then write the answers to the remaining questions. The first one has been done for you.

> Note: We use $0.\dot{1}$ to represent 0.11111 ..., in which 1 is repeated forever after the decimal point. A decimal number in which a digit or a sequence of digits in the decimal part repeats forever is called a **recurring decimal**.

(a) $1 \div 9 = 0.\dot{1}$

(b) $2 \div 9 = \boxed{}$

(c) $3 \div 9 = \boxed{}$

(d) $4 \div 9 = \boxed{}$

(e) $5 \div 9 = \boxed{}$

(f) $6 \div 9 = \boxed{}$

(g) $7 \div 9 = \boxed{}$

(h) $8 \div 9 = \boxed{}$

4 Solve these problems.

(a) A spaceship flew in a circular orbit. One round of the circular orbit is about 42 371 km long. It took the spaceship 90 minutes to fly once round the orbit. How many kilometres did the spaceship fly per second? (Use a calculator.)

(b) According to available statistics, an untightened tap leaks about 0.018 tonnes of water in a day.

(i) Based on the statistics, how much water is wasted in one year (take one year as 365 days) from an untightened tap?

(ii) If that amount of water is poured into drinking water tanks, each with a capacity of about 19 kg of water, about how many tanks can it fill up?

(iii) Suppose that every household uses 3 tanks of water each month. For how many months can the amount of water be used? About how many years is this equivalent to?

Challenge and extension question

5 Use a calculator to find the answers to the first four questions below. Find the pattern and then write the answers to the remaining questions.

(a) 12 345 679 × 1 × 9 =

(b) 12 345 679 × 2 × 9 =

(c) 12 345 679 × 3 × 9 =

(d) 12 345 679 × 4 × 9 =

(e) 12 345 679 × 5 × 9 =

(f) 12 345 679 × 6 × 9 =

(g) 12 345 679 × 7 × 9 =

(h) 12 345 679 × 8 × 9 =

(i) 12 345 679 × 9 × 9 =

2.10 Approximation of products and quotients

Learning objective Round decimals to give approximate answers

Basic questions

1 Round each decimal number...

	... to the nearest one	... to the nearest tenth	... to the nearest hundredth
3.409			
16.032			
5.697			
29.993			

2 Use the column method to calculate the answers.

(a) 7.54 × 48
(Round to the nearest one.)

(b) 0.345 × 81
(Round to the nearest hundredth.)

(c) 63.4 ÷ 25
(Round to the nearest tenth.)

(d) 25.11 ÷ 62
(Round to the nearest hundredth.)

3 Solve these problems.

(a) Mary wanted to buy some fruit for a canteen. The price of apples was £1.65 per kg. She bought 18.5 kg of apples. How much did she pay?

(b) The capacity of an oil bottle is 3 litres. There are 39.75 litres of oil. How many oil bottles are needed?

(c) The table shows the exchange rates on one day in March 2018. Use these exchange rates to solve the following questions. You may use a calculator to work out the calculation.

	GBP (£)
US $1	0.6289
Euro €1	0.8372
Chinese Yuan (CNY, ¥1)	0.0996

(i) On a day in March 2018, Anne's aunt sent her a book from the United States. The price of the book was US $12. What was the price in GBP £?

(ii) Anne's mother exchanged 980 Euros to GBP £. How many British pounds did she get?

(iii) How much would 9995 CNY be when exchanged for GBP £?

Challenge and extension questions

4 Calculate the following.

$0.\dot{1} + 0.0\dot{1} + 0.00\dot{1} + 0.000\dot{1} + 0.0000\dot{1} =$

5 In $4 \div 7$, the digit in the tenth position in the quotient after the decimal point is .

The sum of the first 100 digits in the quotient after the decimal point is .

2.11　Practice and exercise (1)

Learning objective Solve problems with mixed operations, including using brackets

Basic questions

1 Use the laws of operations to fill in boxes with a suitable number and the ◯ with an operation sign.

(a) ☐ × 0.84 = ☐ × 7

(b) ☐ × (3.28 × 8) = (12.5 × 8) × ☐

(c) ☐ × (0.125 + 2.5) = ☐ × ☐ ◯ ☐ × 4

2 Use the column method to calculate the answers. Round to the nearest tenth. (Check the answer to the question marked with *.)

(a) 9 × 0.0247　　　　(b) 13.6 ÷ 27　　　　(c) *8.84 ÷ 17

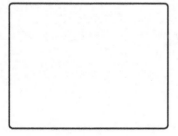 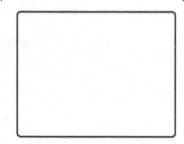 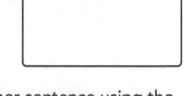

3 Combine the two number sentences into one number sentence using the order of operations.

(a) 60.8 ÷ 16 = 3.8
　　13.5 − 3.8 = 9.7

(b) 4.25 + 5.8 = 10.05
　　14 × 10.05 = 140.7

4 Work these out step by step. (Calculate smartly if possible.)

(a) $9.18 - 9.18 \div 9 \times 3$

(b) $18.7 - 8.7 \div 25$

(c) $19.5 \times 5.8 + 5.2 \times 19.5 - 19.5$

(d) $7.8 \div (39 \div 5)$

(e) $43.2 \div 8 \times 25$

(f) $87.25 - (7.25 + 4.83 + 5.17)$

5 Write the number sentences and then calculate the answers.

(a) 27.9 is divided by the difference between 19.52 and 16.52. What is the result?

_____

(b) Number A is 3.6, which is 4.4 less than 4 times Number B. What is Number B?

_____

6 Solve these problems.

(a) 38 litres of orange juice needs to be poured into bottles, each with a capacity of 750 ml. How many bottles can be filled up? How many litres of orange juice will be left over?

(b) It took a bird half an hour to fly 6.5 km. At this speed, how much time would it take for the bird to fly 13.91 km?

(c) A beekeeper collected 73.8 kg of honey from 36 beehives. Assuming the amount of honey collected from each beehive is the same, how much honey can be collected from 56 beehives?

7 Fill in the spaces to make each statement correct.

(a) All of the digits in a number were moved three places to the right across the decimal point, then moved two places to the left across the decimal point. The resulting number was 4.4.

The original number was ☐.

(b) Put 5.91, 5.9̇ , 5.9, 5.912, and 5.912 912 in order, from the least to the greatest.

The second number is ☐.

(c) Without calculating the answers, fill in the $\bigcirc$ with >, < or =.

78 × 1.4 $\bigcirc$ 78 × 0.4   5.65 ÷ 5 $\bigcirc$ 565 ÷ 50

3.13 ÷ 91 $\bigcirc$ 3.13 ÷ 19   54 $\bigcirc$ 54 × 0.72

97 × 1.5 $\bigcirc$ 97 ÷ 1.5   2.67 ÷ 3 $\bigcirc$ 2.67

(d) When a number with three decimal places is rounded off, it is 4.90. The least possible value of this number is [＿＿＿＿] and the greatest is [＿＿＿＿].

Challenge and extension question

8 The sum of A and B is 6.55. The sum of A and C is 7.55, and the sum of B and C is 7.1. What are A, B, and C, respectively?

A = [＿＿＿＿]   B = [＿＿＿＿]   C = [＿＿＿＿]

Chapter 2 test

1 Calculate mentally and then write the answers.

(a) $5 \times 2.4 = $

(b) $4.7 \div 25 = $

(c) $2.8 + 72 \times 0.1 = $

(d) $9.9 - 4.5 \div 9 = $

2 Use the column method to calculate the answers. Round to the nearest hundredth. (Check the answer to the question marked with *.)

(a) $*10.056 \div 24$

(b) 37×1.28

(c) $7.5 \div 17$

3 Work these out step by step. (Calculate smartly if possible.)

(a) $7.2 \times 25 + 7.2 \times 74 + 7.2$

(b) $8 \times (125 \times 0.125)$

(c) $4.8 \times (90.2 - 13.2)$

(d) $[3.72 - 0.72 \times (1.6 + 2.4)] \div 10$

4 Write the number sentences and then calculate the answers.

(a) Number A is 4.8, which is 0.6 greater than 4 times Number B. What is Number B?

_____

(b) The product of 15 and 0.24 is divided by the sum of 2.85 and 2.15. What is the quotient?

_____

5 Car A and Car B were 450 km apart. They started at the same time and drove towards each other. They met after 2.5 hours. Given that Car A drove at 96 km per hour, find the speed of Car B.

6 A bottle can be filled up with 750 ml of water. There are 9.6 litres of water. How many bottles can it fill up? How much water will be left over?

7 A bottle of olive oil weighed 8 kg, including the bottle itself. After half of the oil was used, it weighed 4.5 kg including the bottle. What was the weight of the bottle?

8 The length of a rectangular lawn is 13.5 m. Its perimeter is equal to the perimeter of a square with side length of 9 m. How many times the width is the length of the rectangular lawn?

9 One room in Sandy's house is 4.2 m long and 4 m wide. Sandy plans to tile the floor with square tiles of 20 cm × 20 cm. How many of the tiles are needed?

For questions 10 to 13, fill in the boxes.

10 Based on 6.75 ÷ 54 = 0.125, use reasoning to work out the answers to the following questions.

(a) 67.5 ÷ 54 =

(b) 675 ÷ 54 =

(c) 1.25 × 54 =

(d) 12.5 × 54 =

11 Fill in each ◯ with >, < or =.

(a) 57 × 1.87 ◯ 57

(b) 93 × 0.72 ◯ 93

(c) 7.2 ÷ 20 ◯ 0.72 ÷ 2

12 If all of the digits in a number are moved two places to the left across the decimal point, the resulting number is ⬚ times the original number.

13 If a number with two decimal places is rounded to the nearest tenth, it is 7.0. The greatest possible value of this number is [　　　]. The least possible value of this number is [　　　].

Questions 14 to 17 are multiple choice questions. For each question, choose the correct answer and write the letter in the box.

14 In the four numbers 0.808, $0.\dot{8}0\dot{8}$, $0.8\dot{0}\dot{8}$, and $0.80\dot{8}$, the least number is [　　].

A. 0.808   B. $0.\dot{8}0\dot{8}$ C. $0.8\dot{0}\dot{8}$ D. $0.80\dot{8}$

15 Rounding 0.999 999 to the nearest hundredth, the answer is [　　].

A. 0.99   B. 1.00   C. 9.00   D. 1

16 When the quotient of 783.5 ÷ 26 is rounded to the nearest tenth, it is [　　].

A. 30   B. 30.2   C. 30.13   D. 30.1

17 When a number with two decimal places is rounded off, the result is 6.8. The range of the possible values of the number is [　　].

A. greater than or equal to 6.75 but less than 6.85

B. greater than or equal to 6.754 but less than 6.84

C. greater than or equal to 6.70 but less than 6.85

D. greater than or equal to 6.75 but less than 6.84

Chapter 3 Introduction to algebra

3.1 Using letters to represent numbers (1)

Learning objective Use letters to represent numbers in equations and formulae

Basic questions

1 Write the number that each letter represents.

(a) 3, 6, 9, A, 15

A =

(b) 2, 1, 2, 3, 2, 4, 5, 6, B, 7, 8, 9

B =

(c) 2 + 6 = 3 + M

M =

(d) 15 ÷ 3 = 10 − Y

Y =

(e) 1, 4, 9, 16, 25, 36, 49, C, 81

C =

(f) 1, 5, 2, 10, 3, 15, X, 20, 5, Y, 6

X =   ; Y =

2 Complete each statement.

(a) $m \times 8$ can be simply written as

_____

(b) $x \times 3 \times y$ can be simply written as

_____

(c) $(9 + a) \times 6$ can be simply written as

_____

(d) $n \times 1 + a \div 2$ can be simply written as

_____

(e) $a \times a \times a$ can be simply written as

_____

(f) $b + b + b + b \times b$ can be simply written as

_____

3 Complete each statement. The first one has been done for you

(a) $(a + b) + c = a + (b + c)$

(b) $a(b + c) = ab + \boxed{}$

(c) $a \div b = (a \div c) \div (b \div \boxed{}) \ (b \neq 0, c \neq 0)$

(d) $a - b - c = a - (b \bigcirc c)$

4 Complete each statement using expressions with letters to show the relations between quantities.

(a) In a triangle, if $\angle 1 = a°$ and $\angle 2 = b°$, then $\angle 3 =$ _____.

(b) In an isosceles triangle, if the base angle is $a°$, the size of the vertex angle is _____.

(c) If the perimeter of a square is C, then the side length of the square is _____.

(d) If A represents the unit price, that is, price per item, X represents the quantity, that is, number of items, and C represents the total price of all the items, then $X =$ _____.

(e) If the area of a rectangle is S and the length is a, then the width is _____.

(f) One frog has 1 mouth, 2 eyes and 4 legs. Two frogs have 2 mouths, 4 eyes and 8 legs.

Three frogs have ☐ mouths, ☐ eyes and ☐ legs.

n frogs have ☐ mouths, ☐ eyes and ☐ legs.

5 Complete the table. Use expressions with letters to show the relations between the three quantities.

Speed (m/minute)	Time	Distance
65	t	
v		210
	6	s

Number of sets made per day	Number of working days	Total number of sets made
x		480
	25	x
30	x	

Unit price	Quantity	Total price
8.5	b	
	y	x
a		z

Challenge and extension questions

6 When the sum of three consecutive even numbers is *a*, then the number in the middle is _____, the least number is _____ and the greatest number is _____.

7 In the following column calculations, the same letter represents the same number and different letters represent different numbers. What numbers do *A*, *B*, *C* and *D* represent respectively to make the column expression true? Fill in the boxes below.

$$
\begin{array}{ccc}
 & A & B \\
+ & C & D \\
\hline
 & 9 & 4 \\
\end{array}
\qquad
\begin{array}{ccc}
 & A & B \\
- & C & D \\
\hline
 & 5 & 8 \\
\end{array}
$$

A = _____ *B* = _____ *C* = _____ *D* = _____

3.2 Using letters to represent numbers (2)

Learning objective Use letters to represent numbers in equations and formulae

Basic questions

1 Multiple choice questions. (For each question, choose the correct answer and write the letter in the box.)

(a) a^2 is equal to ☐ .

    A. $a \times 2$ **B.** $a + 2$ **C.** $a \times a$

(b) $2x - x^2$ is ☐ .

    A. greater than 0   **B.** less than 0

    C. equal to 0   **D.** not sure

(c) Laila is younger than Helen. Laila is a years old and Helen is b years old. After two years, Laila is ☐ years younger than Helen.

    A. 2   **B.** $b - a$ **C.** $a - b$ **D.** $b - a + 2$

(d) Number A is a, which is b less than 4 times Number B. Number B is ☐ .

    A. $a \div 4 - b$ **B.** $(a - b) \div 4$ **C.** $(a + b) \div 4$

2 Use expressions with letters to represent the relations between quantities. The first one has been done for you.

(a) 100 minus the sum of a and b. $100 - (a + b)$

(b) The quotient of 5 divided by x plus n.   _____

(c) 6 times s minus 2.   _____

(d) Subtract 12 times m from 320.   _____

(e) The sum of 80 and b is multiplied by 5. _____

(f) 6 times the sum of b and 90. _____

3 Write the expressions based on the given conditions.

A toy robot costs 50 pounds, a toy aeroplane costs m pounds, and a toy car costs n pounds.

(a) To buy a toy robot and a toy car costs _____ pounds in total.

(b) To buy a toy aeroplane and 2 toy cars costs _____ pounds in total.

(c) To buy a toy robot, a toy aeroplane and a toy car costs _____ pounds in total.

(d) To buy 2 toy aeroplanes and 3 toy cars costs _____ pounds in total.

(e) A toy aeroplane costs _____ pounds more than a toy car.

4 Use expressions with letters to represent these quantities.

(a) A car has travelled t hours at a speed of 85 km per hour. It has travelled _____ km in total.

(b) Jack spent 6 days reading m pages of a book. He read _____ pages of the book every day on average.

(c) There are 24 basketballs and n footballs. The footballs are _____ fewer than the basketballs.

(d) A shirt costs a pounds, and a pair of trousers costs b pounds. The total cost of buying 3 sets of these clothes is _____ pounds.

5 Look at each number sequence carefully and complete its 5th, 6th and *n*th terms. The first one has been done for you.

(a) 2, 4, 6, 8, __10__ , __12__ , ... , __2n__

(b) 0, 5, 10, 15, _____ , _____ , ... , _____

(c) 0, 1, 2, 3, _____ , _____ , ... , _____

(d) 4, 5, 6, 7, _____ , _____ , ... , _____

(e) 13, 23, 33, 43, _____ , _____ , ... , _____

6 Solve these problems.

(a) On a public holiday, a particular model of mobile phone was selling well in a phone shop. 75 of these mobile phones were sold in the morning and 100 in the afternoon. Given that each costs *a* pounds, what is the total value of the sales in the whole day? How much less was the sales value in the morning than in the afternoon?

_____

(b) A highway construction team was tasked to build an *x* metre long highway. It planned to build *m* metres of highway every day. However, in practice, on each day, the team built 2.5 more metres of highway than planned. In how many days was the task completed?

_____ days

Challenge and extension question

7 The figures below show three identical squares, each containing a different number of circles of the same radius. Look at them carefully and then fill in the table.

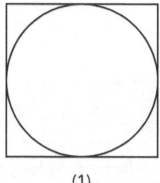

(1)

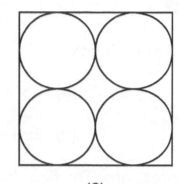

(2)

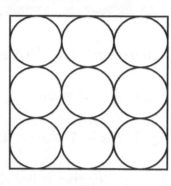

(3)

Figure number	(1)	(2)	(3)	(4)	(5)	(6)	...
Number of circles							

The number of circles in the *n*th square is _____.

Based on the pattern you identified, there are _____ circles in the 2018th figure.

3.3 Simplification and evaluation (1)

Learning objective Simplify expressions and use formulae and equations

Basic questions

1 True or false? Put a ✓ or ✗ in each box.

(a) $6 + a = 6a$ ☐

(b) $n + n - m + m = 2n - 2m$ ☐

(c) $5x + 4 + x = 10x$ ☐

(d) $3x + 4y = 7xy$ ☐

2 Simplify the following expressions. The first three have been done for you.

(a) $5x + 4x = 9x$

(b) $5b + 4b - 9a = 9b - 9a$

(c) $7x + 7 + 6x = 13x + 7$

(d) $5b + 4b - 9b = $ _____

(e) $36s - 15t - 24s + 35t = $ _____

(f) $48x + 75y - 18x - 6x = $ _____

(g) $5 \times 12a = $ _____

(h) $36k \div 9 = $ _____

(i) $6 \times 3x \div 2 = $ _____

(j) $4y \div 2 \times 10 = $ _____

(k) $75x \div 15 + 6 \times 9x = $ _____

(l) $4n \times 7 - 63n \div 9 - 3n = $ _____

3 Complete each statement.

(a) In an equilateral triangle, each side is a metres. Its perimeter

is _____ metres.

(b) Joe has x pencils. Roy has 3 more pencils than Joe. They have

_____ pencils altogether.

(c) Each pack of flour weighs 10 kg. Each pack of rice weighs x kg. y packs

of flour and 5 packs of rice weigh _____ kg in total.

(d) A school bought x boxes of red pens and 10 times as many white pens as red pens.

The school bought _____ boxes of pens altogether.

(e) Don, Evans and Frank each bought 4 pens at a pounds each.

They paid _____ pounds in total for the pens. They also each bought b exercise books. Each book costs 2 pounds.

They paid _____ pounds in total for the books.

4 Solve these problems.

(a) £10 can buy $3a$ kg of pineapples. Fiona bought $9.6a$ kg of pineapples with £50. How much change did she get?

(b) It took Jade m hours to make 21 paper flowers. It took Mariam 2 hours to make n paper flowers.

(i) How many paper flowers did each of them make on average?

    (ii) How many paper flowers did both of them make every hour, on average?

(c) The dividend is 6 times the divisor. If the divisor is x, what is the sum of the dividend, divisor and quotient?

(d) Rob is 4 years older than his younger brother Mike.

    (i) Let x represent Mike's age and y represent Rob's age. Express y in terms of x, and we can get $y = $ _____ .

    (ii) Complete the table to show how old Rob is when Mike is at a certain age.

Mike's age: x	1	2	3	4	...	n	...
Rob's age: y							

Challenge and extension questions

5 Three small rectangles, each with length 5 cm and width a cm, form a large rectangle. What is the area of the large rectangle? What is the perimeter?

6 The length and width of a rectangle are a cm and b cm respectively, and $a > b$. The side length of a square equals the difference between the two sides of the rectangle. What is the sum of their perimeters?

3.4 Simplification and evaluation (2)

Learning objective Solve algebraic problems

Basic questions

1 Using the formula for the nth term for each number sequence below, write its first 5 terms. The first one has been done for you.

nth term	First 5 terms
n	__1__ , __2__ , __3__ , __4__ , __5__
$4n$	_____ , _____ , _____ , _____ , _____
$5n - 3$	_____ , _____ , _____ , _____ , _____
$7 + 7n$	_____ , _____ , _____ , _____ , _____
$\frac{2n}{3} + 2$	_____ , _____ , _____ , _____ , _____

2 Complete the table.

Adam wants to buy a number of books priced at £7 each. The relation between the total price he pays, b, and the number of books he buys, a, can be represented as $7a = b$. When $a = 1, 3, 5, 7, 9, ...$, what values does b represent, respectively?

a	1	3	5	7	9
b					

3 To repair a section of a road, a team repaired c metres of the road every day for the first 6 days, leaving s metres of road still needing repair.

(a) Use an expression to express the length of the road section: _____.

(b) When $c = 50$ and $s = 200$, the length is _____.

4 The side length of a square is a m. Its perimeter is _____ m and its area

   is _____ m². When $a = 5$ m, the perimeter is _____ m and the area

   is _____ m².

5 Simplify first and then evaluate. The first one has been done for you.

   (a) When $x = 2.5$, find the value of $18x - 8x + 6x$.

   $\underline{18x - 8x + 6x = (18 - 8 + 6)x = 16x = 16 \times 2.5 = 40}$

   (b) When $y = 2$, find the value of $12y + 72y \div 6$.

   (c) When $m = 4$ and $n = 1.8$, find the value of $15m + 5m - 18n - 12n$.

   (d) When $a = 0.5$ and $b = 0.6$, find the value of
   $3.5a \times 8 + 75b \div 15 + a - 4b$.

6 Answer these questions using the information given.

   (a) A farm has cultivated sycamores and cedars. They are each in x rows.
   Sycamores are in rows of 12 and cedars are in rows of 14.

   (i) How many sycamores and cedars has the farm cultivated in total?

   (ii) When $x = 20$, how many sycamores and cedars are there on
   the farm?

(b) A car travelled at a speed of a km per hour. It travelled 4 hours in the morning and b km in the afternoon.

(i) Use an expression with letters to represent the distance that the car travelled.

(ii) When $a = 80$ and $b = 200$, what is the distance the car travelled?

Challenge and extension question

7 There are a pupils in a school's athletics team. The number of pupils in the hockey team is 4 fewer than twice the number in the athletics team.

(a) Use an expression with letters to represent the total number of pupils in the two teams. _____

(b) When $a = 24$, how many pupils are there in the two teams?

_____

3.5 Simple equations (1)

Learning objective Solve equations with one unknown

Basic questions

1 True or false? (Put a ✓ for true and a ✗ for false in each box.)

(a) An expression with variables is an equation. ☐

(b) A variable in an equation is also called an unknown. ☐

(c) $9 - 3x$ is not an equation. ☐

(d) $4x - 20 = y$ is not an equation. ☐

(e) $6x \div 2 = 2x + 8$ is an equation. ☐

2 Multiple choice questions. (For each question, choose the correct answer and write the letter in the box.)

(a) In the following, ☐ is not an equation.

    A. $18x + 5x = 23x$ **B.** $5(a + b) = 5a + 5b$

    C. $6x - x - 2x$ **D.** $6y - 8 = 40$

(b) In the following, ☐ is an equation.

    A. $2(a + b)$ **B.** $18 - 2m < 5$

    C. $9 \times 0.9 > 8$ **D.** $x \div 6 = 0$

(c) Given that ▲ + ▲ + ● = 19 and ▲ + ● = 12, then ☐.

    A. ▲ = 9 and ● = 3   **B.** ▲ = 8 and ● = 4

    C. ▲ = 7 and ● = 5   **D.** ▲ = 5 and ● = 7

(d) Given that ● = ▲ + ▲ + ▲ and ● × ▲ = 108,

then ● + ▲ = ☐ .

A. 18   B. 24   C. 54   D. 72

3 Establish an equation with the information given in each diagram below. The first one has been done for you.

m	n
3m	

y	y	y	y
30			

x	8
y	

$m + n = 3m$ _____ = _____   _____ = _____

4 Establish equations based on the relation between equal quantities. The first one has been done for you.

(a) 2 times x equals 36. _Answer: 2x = 36_

(b) The difference between 45 and x is 15. _____

(c) 12 more than 3 times x is 72. _____

(d) 48 is 2 times the sum of x and 3. _____

(e) Half of y is 25. _____

(f) 20 divided by 10 and then plus x is 8. _____

(g) 2 times x is 1 more than 3 times 5. _____

(h) 15 multiplied by 9 is 5 less than 4 times y. _____

5 The sum of Tim's age and his sister Jane's age is 10. Let x be Tim's age and y be Jane's age.

(a) Establish an equation with x and y. _____

(b) Complete the table below to list all the possible combinations of x and y that satisfy the equation. (Note: x and y must both be positive integers.)

x	1	2	3	4	5	6	7	8	9
y									

(c) If Tim is 6 years older than Jane, how old are they?

Tim = _____ Jane = _____

Challenge and extension question

6 Look at the figure. The side length of the larger squares are a cm, and the side length of the smaller squares are b cm. Use letter expressions to represent the relation between a and b. What is the perimeter and the area of the whole figure? If a is 6 cm, find the length of b, the perimeter and the area of the whole figure.

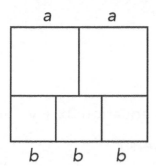

3.6 Simple equations (2)

Learning objective Solve equations with one unknown

Basic questions

1 Complete each statement.

(a) A _____ to an equation is a value we can put in place of the unknown that makes the equation true.

(b) The process to find the _____ to an equation is called 'solving the equation'.

(c) The solution to the equation $2y = 30$ is _____.

(d) The solution to the equation $x + 1.2 + 2.4 = 5$ is _____.

2 Multiple choice questions. (For each question, choose the correct answer and write the letter in the box.)

(a) $x = 7$ is the solution to equation ☐.

    A. $x + 5 = 10$ **B.** $5 - x = 2$

    C. $3 + y = 10$ **D.** $37 - x = 30$

(b) The pairs of numbers for x and y that satisfy the equation $3x + y = 4$ are ☐. (Choose all the correct answers.)

    A. $x = 1, y = 1$ **B.** $x = 2, y = 0$

    C. $x = 1.33, y = 0$ **D.** $x = 0, y = 4$

(c) In solving the equation $50 \div x = 0.5$, the correct working shown below is $\boxed{}$.

A. $50 \div x = 0.5$
Solution: $x = 0.5 \div 50$
$x = 0.01$

B. $50 \div x = 0.5$
Solution: $= 50 \div 0.5$
$= 100$

C. $50 \div x = 0.5$
Solution: $x = 50 \div 0.5$
$x = 100$

D. $50 \div x = 0.5$
Solution: $x = 0.5 \times 50$
$x = 25$

3 Solve the equations and then check your answers. The first one has been done for you.

(a) $x + 8.25 = 11.39$
Solution: $x = 11.39 - 8.25$
$= 3.14$

Check: $3.14 + 8.25 = 11.39$

(b) $72.6 - x = 36.8$

(c) $x - 2.4 = 2.4$

(d) $0.25x = 0.4$

(e) $x \div 3.6 = 0$

(f) $x \div 2.6 = 2.6$

4 Write an equation for each picture. Find the solution.

(a)

I am 28 years older than you.

Luke is x years old. Dad is 40.

(b)

152 cm y cm

I am 5 cm shorter than you.

(c)

I ran 2.8 km this week.

If Eve ran the same distance every day, she ran s m every day.

(d)

a sweets

Shared equally between 25 children, 3 sweets each.

(a) _____ Solution: _____

(b) _____ Solution: _____

(c) _____ Solution: _____

(d) _____ Solution: _____

5 Emily was given £30 in £2 coins and £5 notes. Let x be the number of £2 coins and y be the number of £5 notes.

(a) Establish an equation with x and y.

_____

(b) List all the possible combinations of x and y that satisfy the equation. (Note: x and y must be positive integers.)

Challenge and extension question

6 A department store received 324 pairs of sports shoes packed in boxes with the same number in each. If 2 boxes contain 72 pairs of the shoes, how many boxes did the store receive?

3.7 Simple equations (3)

Learning objective Solve equations with one unknown

Basic questions

1 Fill in the answers.

(a) $16 + x = 44$, solution: _____

(b) $12 \div x = 3$, solution: _____

(c) $6x = 108$, solution: _____

(d) $2x = 3 - x$, solution: _____

(e) $7x + 9x - 15x =$ _____

(f) $16x \div 4 \times 3 - 5x =$ _____

2 Solve the equations. (Check the answers to the questions marked with *.)

(a) *$4x + 2.4 = 16.4$
Solution: $4x = 16.4 - 2.4$
$4x = 14$
$x = 14 \div 4$
$= 3.5$
Check: $3.5 \times 4 + 2.4 = 16.4$

(b) *$7.8 - x \div 3 = 2.2$

(c) $7.2 - 0.3x = 1.8 + 0.3x$

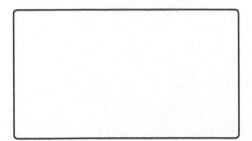

(d) *$(6.8 + 1.2) \div x = 0.8$

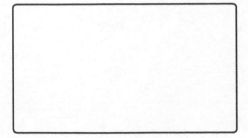

(e) $18(3 + x) = 144$ (f) *$(x + 8) \div 0.5 = 20$

3 Write the equation and then find the solution.

(a) The sum of 4 times x and 3.2 is 9.8. Find x.

_____

(b) 5 times the difference of 12 and x is 40. Find x.

_____

(c) 102 less than 3 times x is 78. Find x.

_____

4 Write the equations based on the information given below and then find the solutions.

(a) Leila bought x books from a shop with £50. Each book cost £8. She had £2 left. Find x.

(b) A canteen bought 3 baskets of carrots with x kg in each basket. 40 kg of carrots was used, which left 38 kg of carrots. Find x.

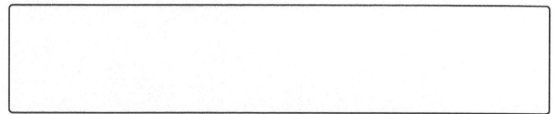

(c) Mr Singh bought 3 e-readers at x pounds each. He paid £200 and got £8 change back. Find x.

Challenge and extension question

5 Write an equation based on the diagram and use reasoning to find x and y.

x	x	x	x
y	y	y	
x	4.5		

Equation: _____ $x =$ _____

Equation: _____ $y =$ _____

3.8 Using equations to solve problems (1)

Learning objective Solve equations with one unknown

Basic questions

1 Solve the equations.

(a) $x \div 5 - 3 = 8$

(b) $6x + 1.5 = 4.5$

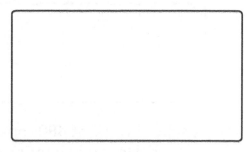

(c) $3x + 4.7 - x = 7.4$

(d) $2(x - 1.5) + x = 12$

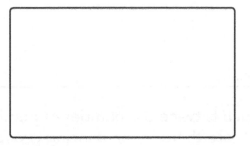

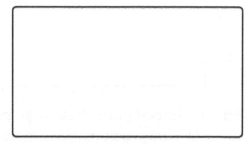

2 Complete the equations on the relations between quantities.

(a) Unit price of pencils × _____ = Total price of pencils

(b) _____ − Number of pear trees = Number of peach trees more than that of pear trees

(c) Amount of work ÷ _____ = Work efficiency

(d) Number of pupils in Class A + _____ = Total number of pupils in Class A and Class B

3 Write equations and then solve the application problems. The first one has been done for you.

(a) There were 105 cars parked in a car park. After some cars were driven away, there were 34 cars left. How many cars were driven away?

> If x is the number of cars that were driven away
> then, $x + 34 = 105$
> $x = 105 - 34 = 71$
> therefore, 71 cars were driven away.

(b) Tom has read 98 pages of a book in two days. He read 55 pages on the first day. How many pages did he read on the second day?

(c) A school bought 480 ropes. They were shared equally among 32 classes. How many ropes did each class receive?

(d) A school choir has 64 pupils, which is twice the number of pupils in the dancing group. How many pupils are there in the dancing group?

(e) Amy can type 86 words in one minute, which is 8 words more than Ayesha can type in one minute. How many words can Ayesha type in one minute?

(f) The area of a rectangle is 36 cm². The length is 8 cm. What is the width of the rectangle?

(g) Mr Brown bought 1 football and 6 volleyballs for £112 in total. The volleyballs were priced at £15 each. How much did the football cost?

Challenge and extension questions

4 Jamelia bought 1 bath towel and 4 hand towels for £20. Given that the price of the bath towel was £6, how much did each hand towel cost?

5 There are two pieces of purple ribbon. The first piece is 4.5 m long. If the second piece has 3.5 m cut from it, the length of the remaining part is 1.2 times the length of the first piece. How long is the second piece?

3.9 Using equations to solve problems (2)

Learning objective Use equations to solve problems

Basic questions

1 Solve these equations.

(a) $7x - 3.5 = 9.1$

(b) $6x = 180 \div 4$

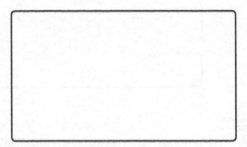

(c) $3(x + 2.3) = 9.6$

(d) $10x \div 2 - x = 6.4$

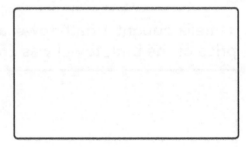

2 Write equations and then solve the problems.

(a) In a primary school, the number of pupils in the chess team multiplied by 4 and then added to 12 is equal to the number of the pupils in the tennis team. There are 80 pupils in the tennis team. How many pupils are there in the chess team?

(b) A box of apples weighs 18 kg, which is 3 kg less than 1.5 times the mass of a box of oranges. What is the mass of a box of oranges?

(c) A pen costs £0.30 more than 3 ballpoint pens. The pen is £1.50. What is the price of a ballpoint pen?

(d) The length of the River Severn is 354 km, which is 338 km shorter than twice the length of the River Thames. How long is the River Thames?

(e) A pair of trainers is priced at £78, which is £14 more than twice the price of a pair of sandals. What is the price of a pair of sandals?

(f) There are some people in a sports field doing physical exercise. 28 people are playing football. If 8 more people join the footballers, the number of people playing football will be 3 times the number of people running. How many people are running?

(g) A new mobile phone shop sold 375 mobile phones in the first month, which was 25 more than half of all the mobile phones in stock. How many mobile phones did the shop have in stock?

Challenge and extension question

3 There are 9 boxes of eggs, each with the same mass. If 15 kg of eggs are taken from each box, the mass of the remaining eggs in the 9 boxes is equal to the original mass of 4 boxes. What was the original mass of each box of eggs?

3.10 Using equations to solve problems (3)

Learning objective Use equations to solve problems

Basic questions

1 Solve these equations.

(a) $90 - 5x = 35$

(b) $1.8x + 1.5x + 3.4 = 10$

(c) $(9x + 27) \div 2 = 81$

(d) $3x - 2(x + 1) = 8$

2 Write equations and then solve the problems.

(a) A canteen received a delivery of 150 kg of rice, which was 30 kg less than 3 times the flour it received at the same time. What was the mass of the flour the canteen received?

(b) Warehouse A stored 56 tonnes of food. The amount was 8 tonnes more than twice the food stored in Warehouse B. How many tonnes of food did Warehouse B store?

(c) Jack's mum is 35 years old. This is 2 years more than 3 times Jack's age. How old is Jack?

(d) 36 white rabbits and some grey rabbits are on the grass. There are 3 times as many white rabbits as grey rabbits. How many grey rabbits are on the grass?

(e) The volume of a big bottle is 2.5 litres, which is 250 ml less than 5 times the volume of a small bottle. What is the volume of the small bottle?

(f) Dad's age is 4 times the age of his son. Dad is 48 years older than his son. How old are the son and dad?

(g) A tailor bought 72 m of cloth, which was exactly enough to make 20 sets of adult clothes and 16 sets of children's clothes. 2.4 m of cloth is needed for each set of adult clothes. What is the length of cloth needed for each set of children's clothes?

(h) Three years ago, a mum's age was 6 times the age of her daughter. The mum is 33 years old this year. How old is her daughter this year?

Challenge and extension question

3 A farm had the same number of goats and sheep. After the farmer sold 200 goats and bought 300 sheep, there were 6 times as many sheep as goats. How many goats and sheep were there on the farm at first?

Chapter 3 test

1 Solve the equations. (Check the answers to the questions marked with *.)

(a) $x - 2.3 = 2.7$

(b) $1.8 + 3x = 9.6$

(c) $25.2 - x \div 2 = 8$

(d) $*4x + 1.2 \times 5 = 24.4$

(e) $*0.4(5x + 2) = 2.8$

(f) $2x - 8.6 + 0.5x = 5.4$

2 Solve the following evaluation questions. (Simplify first if possible.)

(a) When $a = 4$, $b = 5$ and $c = 6$, find the value of $bc - ac$.

(b) When $a = 7$ and $b = 2.5$, find the value of $5a + 4b - (4a - 3b)$.

(c) Using the formula for the nth term for each number sequence below, write its first 5 terms.

nth term	First 5 terms
$2n - 1$	_____ , _____ , _____ , _____ , _____
$3n + 2$	_____ , _____ , _____ , _____ , _____
$7(n - 1)$	_____ , _____ , _____ , _____ , _____
$\frac{n}{2} + 1$	_____ , _____ , _____ , _____ , _____

3 Write the equations and then find the solutions.

(a) The sum of 2.5 times a number and 5 is 25. Find the number.

(b) 6.5 times a number minus 4 times the number is 25. Find the number.

(c) 8 times the difference of a number and 5 and then divided by 3 is 120. Find the number.

For questions 4 to 9, use equations to solve the problems.

4 A canteen bought 8 kg of tomatoes. It paid £35 pounds and got £1.40 change back. What was the price of 1 kg of tomatoes?

5 Jay has 55 science books and some storybooks. The number of science books is 14 fewer than 3 times the number of storybooks. How many storybooks does Jay have?

6 A road construction team was paving a road. They paved 0.4 km each day. After 8 days, they completed 0.5 km more than half the length of the road. What is the total length of the road?

7 A computer manufacturer planned to assemble 5800 computers in 20 days. 440 more computers than planned were assembled in 20 days. How many computers were actually assembled each day on average?

8 A company has 800 staff members. The number of female staff members is 40 fewer than twice the number of male staff members. How many male and female staff members are there in the company?

9 There were 40 tonnes of water in total in two ponds. After 4 tonnes of water were poured into Pond A, and 8 tonnes of water were discharged from Pond B, the amount of water in both ponds was the same. How many tonnes of water did the two ponds have at first?

For questions 10 to 17, fill in the spaces to complete each statement.

10 A road construction team repaired 2.4 km of a road in x days. On average, this was _____ km per day.

11 James has a stamps, which is 3 stamps fewer than Simon. Simon has _____ stamps.

12 Class A in Year 6 has a pupils. 3 pupils were absent one day. _____ pupils were present.

13 An orchard has x pear trees. The number of apple trees is 10 more than twice the number of pear trees. The orchard has _____ apple trees.

14 If the unit price for 1 kg of apples is a pounds, then buying 3 kg of the apples will cost _____ pounds. To buy another 5 kg of the apples, the total cost will be _____.

15 A school has x boys, which is 24 more than the number of girls.

There are _____ pupils in total in the school. When $x = 652$, there

are _____ pupils in total.

16 If the perimeter of a square is $2a$ cm, then its area is [____] cm².

17 James, Tim and Gary had a 100-metre race. It took James x seconds to finish the race. It took Tim 2 more seconds than James and it took Gary 0.3 seconds less than Tim to finish the race.

_____ was the champion.

For questions 18 to 22, determine whether the statement is true or false. (Put a ✓ for true and a ✗ for false.)

18 Both t^2 and $2t$ represent t times t. [ ]

19 An expression is not an equation. [ ]

20 Simplify: $5x - 4x - 9 + 8 = x - 17$. [ ]

21 A value of the unknown which makes an equation true is called a solution

to the equation. [ ]

22 Three people went to buy sports goods with the same amount of money. Person A bought 3 footballs, Person B bought 4 basketballs, and Person C bought 1 football, 1 basketball and 2 volleyballs. If each football costs

$4x$ pounds, then each volleyball costs $2.5x$ pounds. [ ]

Multiple choice questions. (For questions 23 to 28, choose the correct answer and write the letter in the box.)

23 In the following expressions, [ ] is an equation.

    A. $25x$ **B.** $15 - 3 = 12$ **C.** $6x + 1 = 6$ **D.** $4x + 7 < 9$

24 $x = 3$ is the solution to equation ☐.

    A. $3x = 4.5$ **B.** $2x + 9 = 5x$ **C.** $1.2 \div x = 4$ **D.** $3x \div 2 = 18$

25 Two pairs of numbers for x and y that both satisfy the equation

    $2x + 3y = 10$ are ☐.

    A. $x = 5, y = 0$ and $x = 0, y = 5$ **B.** $x = 2, y = 2$ and $x = 1, y = 3$

    C. $x = 3.5, y = 1$ and $x = 4, y = \frac{2}{3}$ **D.** $x = 0.5, y = 3$ and $x = 1.5, y = 4$

26 5 squares with side length a are put together to form a large rectangle.

    The perimeter of the large rectangle is ☐.

    A. $5a$ **B.** $10a$ **C.** $12a$ **D.** $20a$

27 Year 6 pupils planted a trees, which was b trees fewer than twice the number of trees planted by Year 5 pupils. Year 5 pupils planted ☐ trees.

    A. $(2a - b)$ **B.** $(a + b) \div 2$ **C.** $2(a + b)$ **D.** $(a - b) \div 2$

28 If $2a + 2b + 1 = 6$, then the result of $5(a + b) - 4$ is ☐.

    A. 2.5 **B.** 4.5 **C.** 8.5 **D.** 12.5

Chapter 4   Geometry and measurement (I)

4.1 Perpendicular lines

Learning objective   Identify perpendicular lines in shapes and intersecting lines

Basic questions

1 Complete each statement. (For questions (a) and (b), choose from 'perpendicular', 'perpendicular foot', 'angles' and 'right angles'.)

(a) When two straight lines intersect at a right angle, these two lines

are _____ to each other, and one of the lines is

called a _____ line to the other line. The point

where the two lines intersect is called the _____.

(b) When two lines intersect they form four _____.
If one of the angles is a right angle, then the other three angles are

also _____ , and the two lines are

_____ to each other.

(c) In the figure on the right, *AB* and *CD* are _____
to each other, denoted as

AB ⊥ _____ or

CD ⊥ _____ .

They are read as *AB*

is _____ to *CD*

or *CD* is _____ to *AB*.

(d) The two adjacent sides of a rectangle are _____ to each other.

(e) At _____ o'clock, the hour hand and the minute hand on a clock face are perpendicular to each other.

(f) When two lines are perpendicular to each other, all the angles they form are _____ .

2 Use a set square to measure and then find whether each pair of lines are perpendicular to each other. If they are perpendicular, mark them in the figure with the right angle sign and denote them with letters and the perpendicular sign '⊥' in the boxes below.

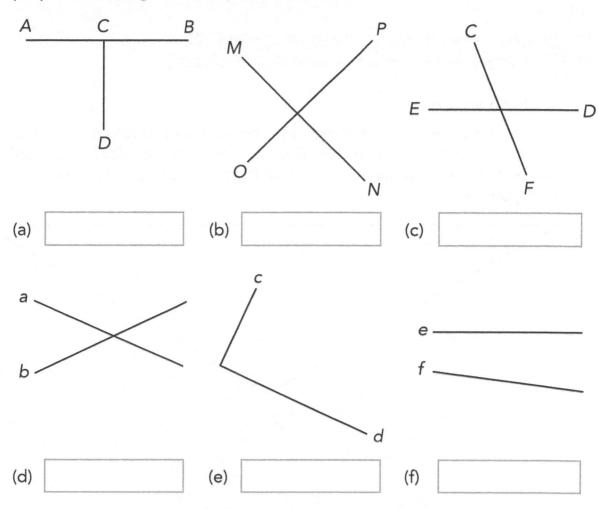

(a) [_____] (b) [_____] (c) [_____]

(d) [_____] (e) [_____] (f) [_____]

3 Measure and complete each statement.

The figure on the right shows four line segments PA, PQ, PB and PC. A, Q, B and C are all on the line *l* and P is a point outside the line *l*. The shortest line segment is [   ] , and it is the _____ from point P to the line *l*.

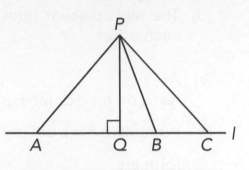

A **line segment** (or simply a segment) is part of a line with two endpoints.

Challenge and extension question

4 Hands-on activity.

The figure below shows an isosceles triangle. Can you fold it to form a right angle? Is it possible to fold it to form two right angles? How many right angles can you fold at the most? Try it yourself. Draw in lines to show your results.

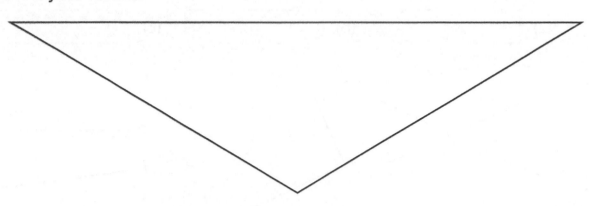

4.2 Parallel lines

Learning objective Identify parallel lines in shapes

Basic questions

1 Look at the figures. Which two lines are parallel? Circle the letter(s) to show your answer.

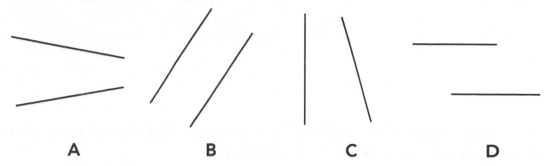

| A | B | C | D |

2 True or false? (Put a ✓ for true and a ✗ for false in each box.)

(a) Two parallel lines never intersect. ☐

(b) Two lines are either parallel or perpendicular. ☐

(c) If two lines *a* and *b* are parallel, we write *a* // *b*. ☐

(d) A rectangle has two pairs of parallel lines and four pairs of perpendicular lines. ☐

(e) In the figure below, line *b* and line *c* are parallel. ☐

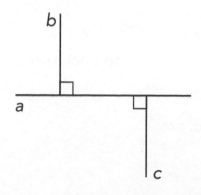

3 Use a set square to find the parallel lines in each figure below.
Denote these parallel lines with // in the boxes.

(a)

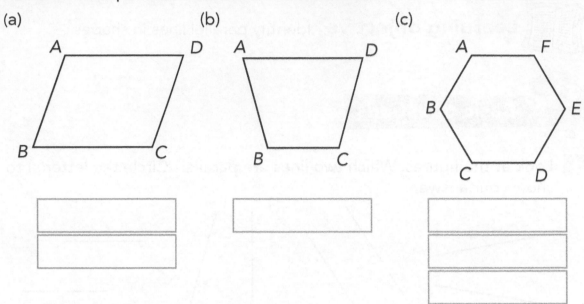

(b)

(c)

4 Look at the image of the ladder. Think carefully and then choose from the following terms to complete the statements:

> vertical   horizontal   parallel   perpendicular
> a // b $a \perp b$ c // d // e
> a is parallel to b a is perpendicular to b equal

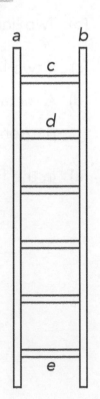

(a) In the ladder shown on the right, the two sides a

and b are both _____.

a is _____ to b. It can be

denoted as _____ and read as

_____.

(b) The steps c, d and e are all

_____ and they

are _____ to each other.

It can be denoted as _____.

The lengths of them are

all _____.

5 How many pairs of parallel sides are there in a square, in a regular hexagon, and in a regular octagon? Can you also tell how many pairs of parallel sides are in a regular *n*-gon, that is, a regular polygon with *n* sides, when *n* is an even number? Fill in the table.

Figure	square	regular hexagon	regular octagon	regular *n*-gon
Number of pairs of parallel sides				

4.3 Parallelograms (1)

Learning objective Recognise and use the properties of parallelograms

Basic questions

1 Complete each statement.

(a) In the figure on the right,

_____ // _____

and _____ // _____ .

A quadrilateral like this in which two pairs of opposite sides

are _____ is called a _____ denoted as

▱ ABCD. AC and BD are called _____ of ▱ ABCD.

(b) The opposite sides of a parallelogram are parallel to each other, and

their lengths are _____. The opposite angles are

also _____.

(c) A parallelogram with one angle that is a right angle is

a _____.

(d) A rectangle with all four sides of equal length is a _____.

(e) Both _____ and _____ are special
parallelograms.

(f) In the following figures, the parallelogram(s)

is/are _____.

2 True or false? (Put a ✓ for true and a ✗ for false in each box.)

(a) Any parallelogram can be cut into two identical triangles along its diagonal. ☐

(b) Any two sides of a parallelogram are parallel to each other. ☐

(c) The opposite angles of a parallelogram are equal, and they are right angles. ☐

(d) A quadrilateral with all four sides of equal length is called a square. ☐

(e) The total length of the four sides of a parallelogram is its perimeter. ☐

(f) A parallelogram is a symmetrical figure over a symmetry line. ☐

3 The figure below shows part of each of three parallelograms on a square grid. Draw lines to complete the parallelograms.

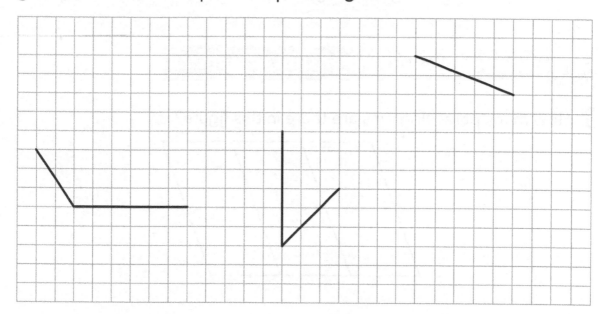

4 Multiple choice questions. (For each question, choose the correct answer and write the letter in the box.)

(a) Cut a parallelogram into two pieces and then put them together.

It is definitely possible for the new figure formed to be a [   ].
(Choose all the correct answers.)

A. rectangle   B. square

C. triangle   D. parallelogram

(b) Cut a parallelogram along its two diagonals to get four triangles. It is impossible that they are [   ].

A. four acute-angled triangles

B. four obtuse-angled triangles

C. two right-angled triangles and two obtuse-angled triangles

D. two acute-angled triangles and two obtuse-angled triangles

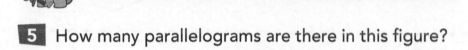

Challenge and extension questions

5 How many parallelograms are there in this figure?

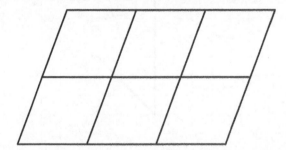

6　There are six small sticks. Their lengths are 1 cm, 2 cm, 3 cm, 5 cm, 6 cm and 7 cm, respectively. To use these six small sticks to form a parallelogram, the opposite sides of the parallelogram will be _____ and _____, or _____ and _____.

4.4 Parallelograms (2)

Learning objective Recognise and use the properties of parallelograms

Basic questions

1 Complete each statement.

(a) In the figure, *MN* is the _____ of parallelogram *ABCD*, and side *BC* is called the _____ of the parallelogram.

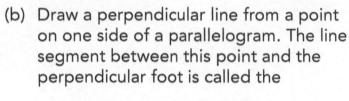

(b) Draw a perpendicular line from a point on one side of a parallelogram. The line segment between this point and the perpendicular foot is called the

_____ on the base of the parallelogram.

(c) The figure shows a parallelogram *ABCD*. If the base is *AB*, then the height

is _____. If the height is *AE*,

then the base is _____

or _____ .

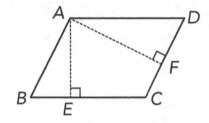

(d) The lengths of all the heights on the same base of a parallelogram

are _____.

(e) The figure on the right shows a parallelogram. Its height shown in

the figure should be _____

or _____ .

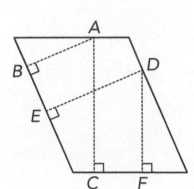

(f) Passing through a point on the base of a parallelogram, at

most _____ height(s) can be drawn.

(g) If two pairs of opposite sides of a quadrilateral are parallel to each other, and one of the angles is a right angle, then the

quadrilateral is called a _____; it is also called a

special _____.

2 Multiple choice questions. (For each question, choose the correct answer and write the letter in the box.)

(a) The base of parallelogram *ABCD* is *AB*,

and its height is ☐.

A. *CD*   B. *AE*

C. *DF*   D. *BC*

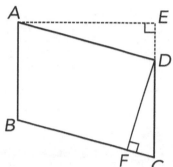

(b) The figure shows a parallelogram *ABCD*.

For ☐, the side is not the base for the height.

A. *AB* and *CE*   B. *BC* and *MN*

C. *AD* and *CD*   D. *CD* and *CE*

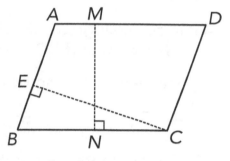

(c) On one side of a parallelogram, you can draw ☐ heights.

A. 1   B. 2

C. 4   D. infinitely many

3 Draw three different parallelograms in the 1 cm square grid below, each with a base of 3 cm and a height of 2 cm.

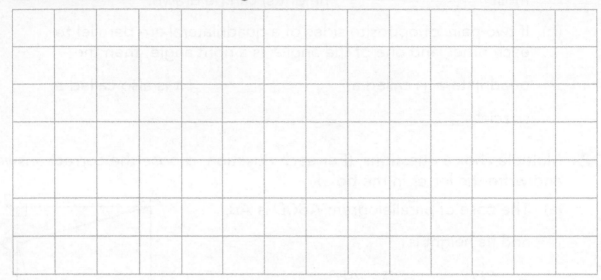

Challenge and extension questions

4 In parallelogram *ABCD* shown on the right, *AB* is 6 cm and *BC* is 8 cm.

The height *AE* can be [ ] cm.

A. 5   **B.** 6   **C.** 7

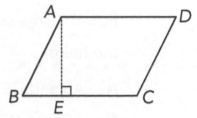

5 The perimeter of parallelogram *ABCD* is 50 cm, *AB* is 15 cm, and then *BC* is [ ] cm.

6 How many parallelograms are there in the figure on the right?

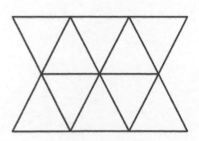

4.5 Area of a parallelogram

Learning objective Calculate the area of any parallelogram

Basic questions

1 Complete the table.

Parallelogram	Base	3.2 m	24 cm	18 mm	3 cm
	Height	15 m	70 cm		
	Area			90 mm²	7.5 cm²

2 Calculate the area of each parallelogram. (Unit: cm)

(a)
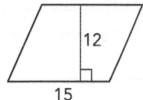

(b)
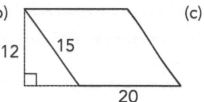

(c)
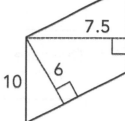

_____

_____

3 Multiple choice questions. (For each question, choose the correct answer and write the letter in the box.)

(a) Look at the figure. To find the area of the parallelogram, the correct calculation is ☐ . (Unit: cm)

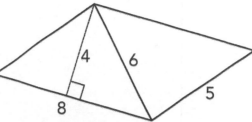

   A. 5 × 6   **B.** 8 × 4

   C. 8 × 5   **D.** 4 × 5

(b) Look at the figure. To find the length of CD, the correct calculation

is ☐ . (Unit: cm)

A. $5 \times 4 \div 3$ **B.** $3 \times 4 \div 5$

C. $5 \times 3 \div 4$ **D.** $5 \times 3 \times 4$

4 Solve these problems.

(a) The area of a parallelogram is $10.8\,m^2$ and its height is $24\,m$. Find the length of the base corresponding to the height.

(b) The base of a parallelogram is $45\,cm$, which is $1.3\,cm$ more than its height. Find the area.

(c) A piece of land is in the shape of parallelogram with a base of $28\,m$ and a height of $15\,m$. The owner plans to cover the land with turf, which costs £4.50 per m^2. What is the total cost to cover the whole area?

(d) The figure shows a parallelogram with a perimeter of 36 cm. AB is 7.8 cm and AE is 5 cm.
Find the area of the parallelogram.

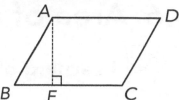

5 The figure shows the shaded parts of two parallelograms with one common base. The relationship between the areas of the shaded parts, A and B, is ☐ .

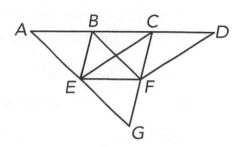

A. $A > B$ **B.** $A = B$ **C.** $A < B$ **D.** not comparable

6 In the figure, EF // AD, BE // CG, BF // AG and CE // DF. Write the parallelograms that have the same areas. (Use letters to represent the parallelograms.)

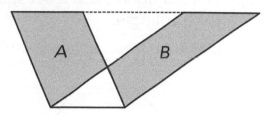

4.6 Area of a triangle (1)

Learning objective Calculate the area of any triangle

Basic questions

1 Complete the table

Triangle	Base	15 m	27 cm	4.4 m	1.25 cm
	Height	12 m	8 cm	15 m	36 cm
	Area				

2 In each figure, use a set square to draw the height to the given base of each triangle.

(a)

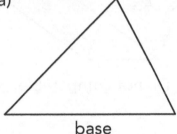

base

(b)

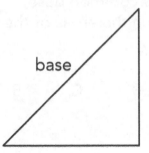

base

(c)

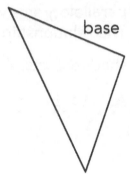

base

3 Complete each statement.

The figure on the right shows △ABC with three sides and three heights. Three pairs of corresponding base and height are

base AC and height _____,

base _____ and height AD,

and base _____ and height _____.

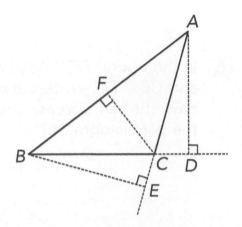

4 Look at each figure and find the area of the triangle. (Unit: cm)

(a)

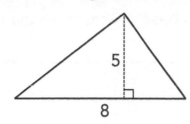

(b)

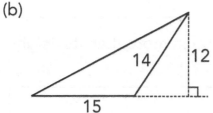

(c)

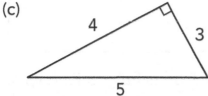

5 True or false? (Put a ✓ for true and a ✗ for false in each box.)

(a) If a triangle and a parallelogram share one common base, then the area of the triangle is half of the area of the parallelogram. ☐

(b) There is only one height in an obtuse triangle. ☐

(c) There are three pairs of corresponding base and height in a triangle. ☐

(d) In a right-angled triangle, if the lengths of the two sides of the right angle are given, then its area can be found. ☐

6 Solve these problems.

(a) The height of a triangle is 24 cm and its base is 3 cm more than three times its height. Find its area.

(b) The base of a triangle is 32 cm, which is 7 cm more than its height. What is its area?

(c) The base of a triangle is 0.18 m, which is $\frac{1}{5}$ of its height. Find its area.

7 In a right-angled triangle ABC, $\angle A = 45°$ and $BC = 6\,cm$. Find its area. (Drawing not to scale.)

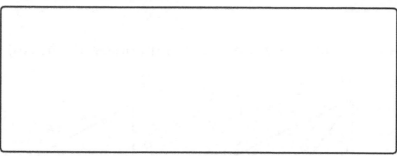

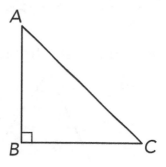

8 In a right-angled triangle, the side lengths are 3.6 cm, 4.8 cm and 6 cm, respectively. Find its area.

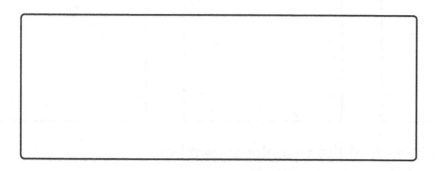

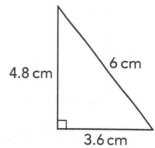

4.7 Area of a triangle (2)

Learning objective Solve problems involving the areas of triangles

Basic questions

1 Look at the figures and calculate the areas of the triangles. (Unit: cm)

(a)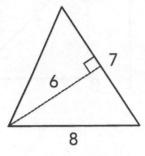

(b)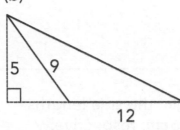

(c)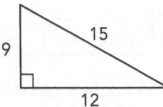

2 Find the unknown in each of the triangles shown below.
(Note: *S* represents the area of a triangle.)

(a) $S = 20\,cm^2$

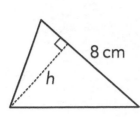

(b) $S = 32.4\,\text{m}^2$

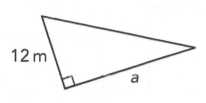

(c) $S = 1536\,\text{cm}^2$

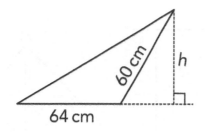

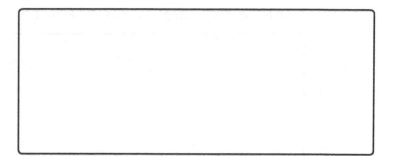

3 Solve these problems.

(a) A flowerbed is shaped like a triangle. The height of this triangle is 18 m which is 8 m less than its corresponding base. What is the area of the triangle?

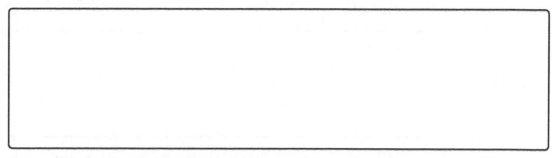

(b) A triangular-shaped vegetable plot has a base of 24 m and a height of 16 m. If the base is increased by 6 m and the height is increased by 3 m, by how much will the area of the plot be increased?

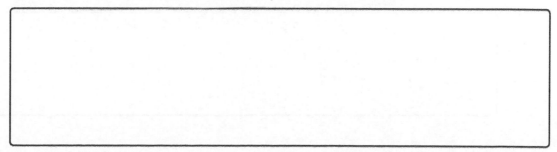

(c) The figure shows a rectangle *ABCD* with an area of 6600 cm². *AB* is 60 cm and *BE* is 40 cm. Find the area of triangle *CDE*.

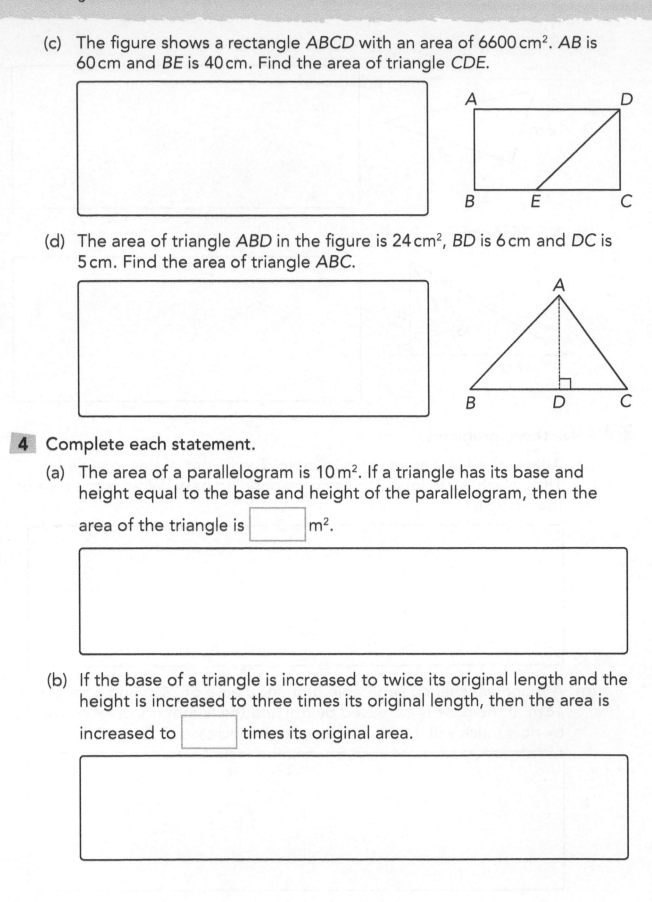

(d) The area of triangle *ABD* in the figure is 24 cm², *BD* is 6 cm and *DC* is 5 cm. Find the area of triangle *ABC*.

4 Complete each statement.

(a) The area of a parallelogram is 10 m². If a triangle has its base and height equal to the base and height of the parallelogram, then the area of the triangle is ☐ m².

(b) If the base of a triangle is increased to twice its original length and the height is increased to three times its original length, then the area is increased to ☐ times its original area.

(c) The area of a triangle is the same as the area of a parallelogram and their heights are also equal. If the base of the parallelogram is 100 cm, then the base of the triangle is ☐ cm.

(d) The figure shows a right-angled triangle ABC. If $AB = 3$ cm, $BC = 4$ cm and $AC = 5$ cm, then $BD = $ ☐ cm.

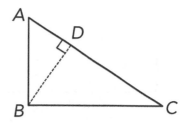

Challenge and extension questions

5 What is the sum of the areas of the shaded parts in the figure?

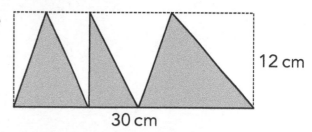

12 cm

30 cm

6 The area of triangle *ABC* in the figure on the right is 48 cm² and the length of *BD* is twice that of *DC*. Find the area of triangle *ABD*.

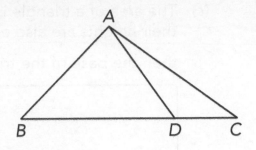

4.8 Practice and exercise (2)

Learning objective Solve problems involving lines and shapes

Basic questions

1 Complete each statement.

(a) When two lines intersect to form a right angle, the two lines

are _____ to each other.

(b) A parallelogram has ☐ pairs of parallel lines.

(c) A triangle has ☐ pairs of parallel lines.

(d) In a rectangle, there are ☐ pairs of sides perpendicular to each

other and ☐ pairs of sides parallel to each other.

(e) A parallelogram and a triangle have the same area and the same
height. The base of the parallelogram is 2a. The base of the triangle

is ☐ .

2 Multiple choice questions. (For each question, choose the correct answer
and write the letter in the box.)

(a) There are ☐ perpendicular line(s) between two parallel lines.

A. 0   **B.** 1

C. 2   **D.** infinitely many

(b) In the following statements, the correct one is ☐.

    A. There are four pairs of parallel lines in a parallelogram.

    B. The four sides of a square are perpendicular to each other.

    C. The four sides of a square are parallel to each other.

    D. Both squares and rectangles are special parallelograms.

(c) In the figure below, $a \parallel b$. The areas of $\triangle ABC$, $\triangle EBC$, and $\triangle FBC$ are denoted by $S_{\triangle ABC}$, $S_{\triangle EBC}$ and $S_{\triangle FBC}$, respectively.

The correct statement is ☐.

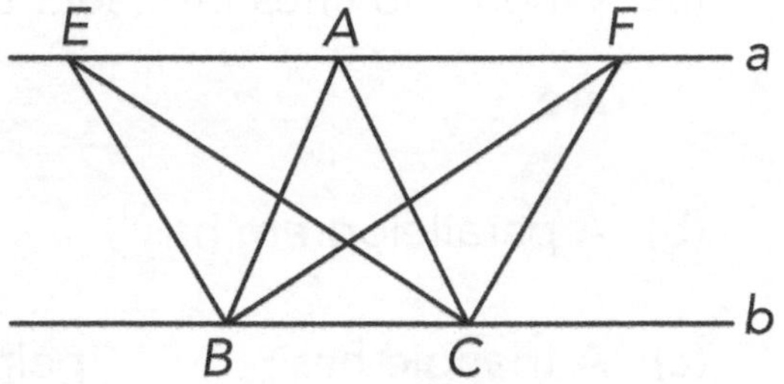

    A. $S_{\triangle ABC}$ is the largest.

    B. $S_{\triangle EBC}$ is the largest.

    C. $S_{\triangle FBC}$ is the largest.

    D. The areas are all the same.

3 The figure shows a piece of rectangular lawn with a length of 40 m and a width of 30 m. There are two paved paths through it. The width of each path is 2 m. Find the area of the lawn.

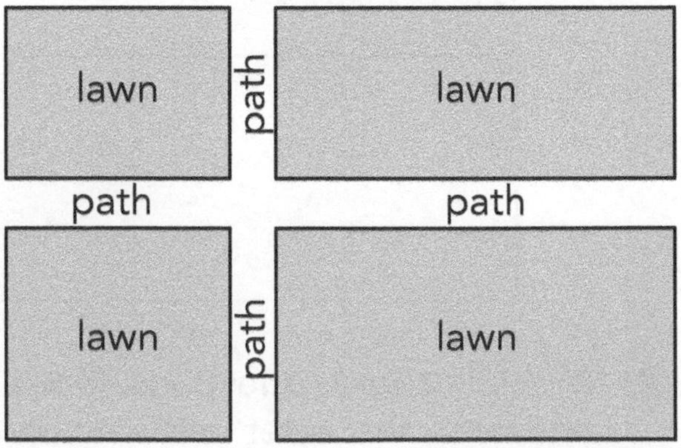

4 To cut out an acute-angled triangle with a base of 3 cm from a rectangle of 4 cm × 6 cm, the greatest possible area of the acute-angled triangle that could be

cut out is ☐ cm².

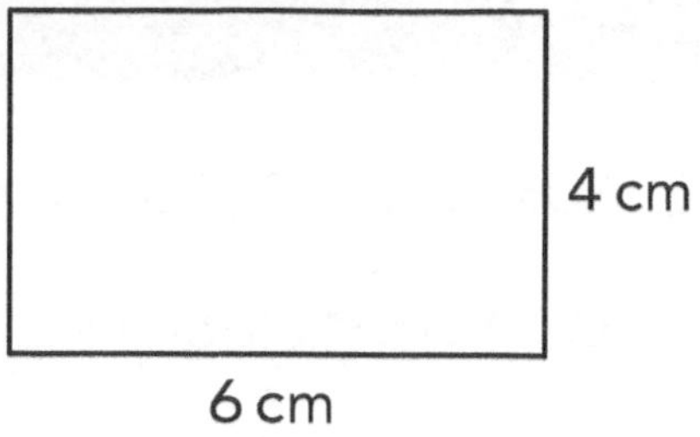

(Hint: the base of the triangle can be horizontal or vertical or in another position. You may use a drawing to help find the answer.)

Challenge and extension question

5 The figure shows a rectangle ABCD. AD is 18 cm and AB is 15 cm. E is the midpoint of BC (that is, BE = EC) and F is the midpoint of CD. Find the area of triangle AEF.

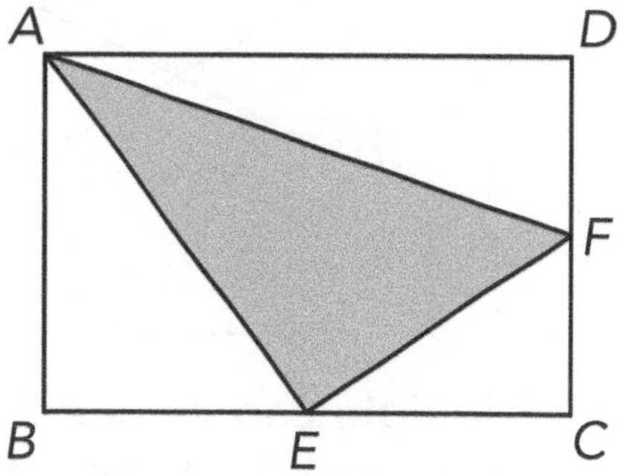

Chapter 4 test

1 True or false? (Put a ✓ for true and a ✗ for false in each box.)

(a) A parallelogram is either a rectangle or a square. ☐

(b) A parallelogram has line symmetry. ☐

(c) The two opposite sides of a square are parallel. ☐

(d) The two adjacent sides of a rectangle are perpendicular to each other. ☐

(e) If the areas of two triangles are equal, then their bases and heights are also equal. ☐

(f) Find the relation between each pair of lines below. If they are neither perpendicular nor parallel, put a ✗ in the box. If they are either perpendicular or parallel to each other, write the letters or the correct sign in the box.

☐ ☐ ☐ ☐

2 Multiple choice questions. (For each question, choose the correct answer and write the letter in the box.)

(a) When the area of a triangle is 24 cm², then its base and height can be ☐ . (Choose all the correct answers.)

    A. 5 cm and 4.8 cm   **B.** 3 cm and 16 mm

    C. 3.2 cm and 15 cm   **D.** 9 cm and 6 cm

(b) When the base and height of a parallelogram are both doubled, its area is increased to ☐ its original area.

    A. 2 times   **B.** 3 times   **C.** 4 times   **D.** 8 times

(c) In the figure on the right, the area of the triangle is $\frac{1}{5}$ of the area of the rectangle. The indicated base of the triangle is ☐.

A. 3 cm   B. 6 cm

C. 9 cm   D. 12 cm

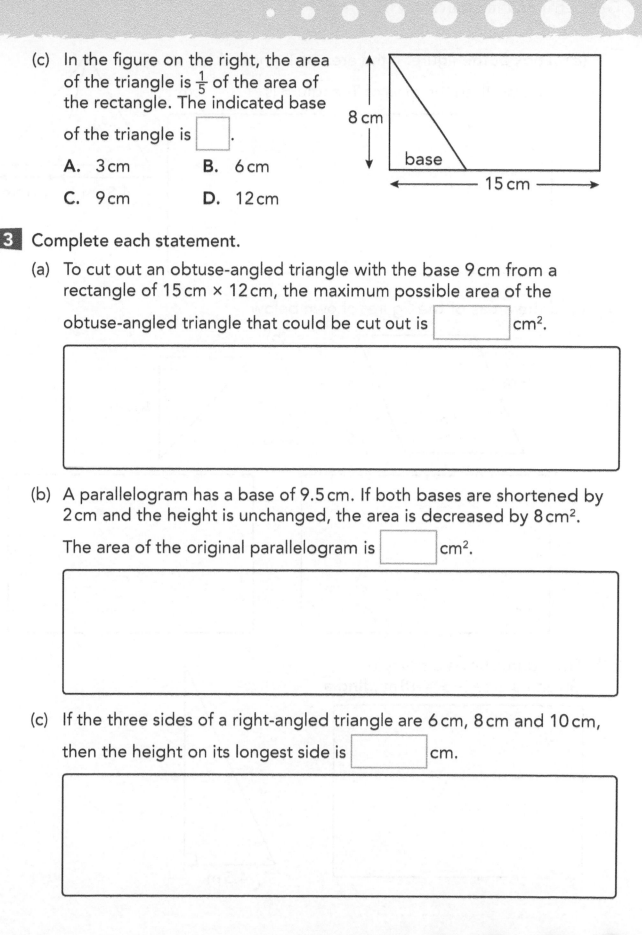

8 cm

base

15 cm

3 Complete each statement.

(a) To cut out an obtuse-angled triangle with the base 9 cm from a rectangle of 15 cm × 12 cm, the maximum possible area of the obtuse-angled triangle that could be cut out is ☐ cm².

(b) A parallelogram has a base of 9.5 cm. If both bases are shortened by 2 cm and the height is unchanged, the area is decreased by 8 cm². The area of the original parallelogram is ☐ cm².

(c) If the three sides of a right-angled triangle are 6 cm, 8 cm and 10 cm, then the height on its longest side is ☐ cm.

(d) Look at the figure. If the area of Triangle A is 12 cm², then the area of Triangle B is [   ] cm².

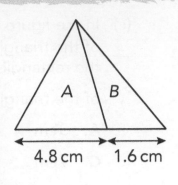

A B

4.8 cm   1.6 cm

4 Find the areas of the figures shown below.

(a)

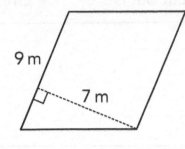

9 m

7 m

(b)

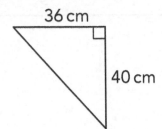

36 cm

40 cm

5 The figure shows a triangle.
Given its area is S = 18 m², find a.

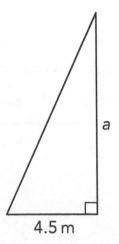

a

4.5 m

6 The figure shows a parallelogram with a perimeter of 66 cm. Find its area.

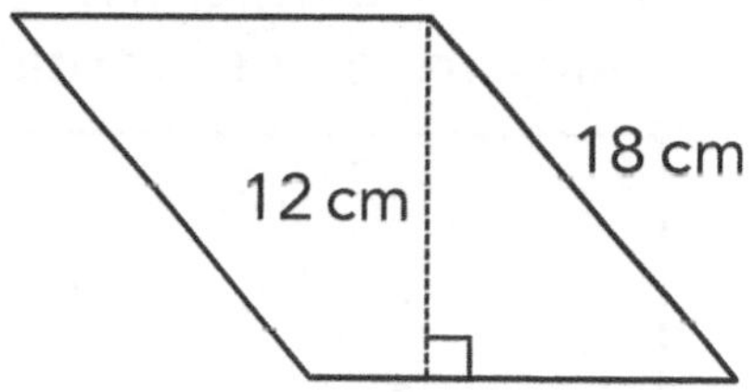

12 cm

18 cm

7 An area of woodland is in the shape of a triangle. The length of its base is 120 m, which is three times the length of its height. Every square metre of the woodland can produce 75.8 grams of oxygen in a day. How many kilograms of oxygen can this woodland produce in a day?

8 The area of a parallelogram is 1.5 times the area of a triangle. The base of the parallelogram is 36 cm, its height is 25 cm, and the base of the triangle is 32 cm. Find the height of the triangle.

9 A piece of rectangular-shaped paper has a length of 1 m and a width of 0.9 m. Amy wants to cut out right-angled triangles from the paper so that in each triangle, the two sides of the right angle are 6 cm and 5 cm. How many triangles in total can Amy cut out from the paper?

10 A triangular-shaped signboard has a base of 1.5 m and the height is 2 m. To paint both sides of the signboard, 3.6 kg of paint is needed. How many kg of paint is needed to cover one square metre of the signboard?

Chapter 5 Consolidation and enhancement

5.1 Operations of decimals (1)

Learning objective Solve calculation problems involving decimal numbers

Basic questions

1 Work these out mentally. Write the answers.

(a) 2.7 + 7.2 = ⬚

(b) 7.1 − 2.9 = ⬚

(c) 5.25 × 4 = ⬚

(d) 40 ÷ 25 = ⬚

(e) 10.6 − 0.6 × 15 = ⬚

(f) 1.4 ÷ 8 × 12 = ⬚

2 Choose the column method to calculate the following.

(a) 24 × 10.5   (b) 4.35 × 328   (c) 49 ÷ 14   (d) 77.55 ÷ 33

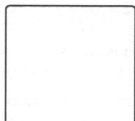

3 Work these out step by step. (Calculate smartly if possible.)

(a) 7.25 ÷ 25 + (4.38 − 2.61)   (b) 1.25 × 56 ÷ 7

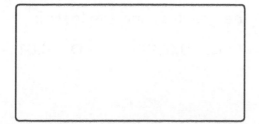
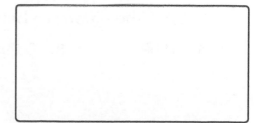

(c) $(172.5 - 72.5) \times 0.12 \div 10$

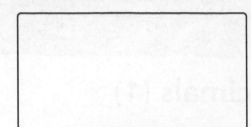

(d) $48 \times 7.6 + 3.4 \times 48 - 48$

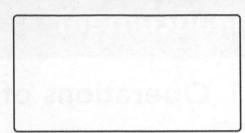

4 Complete each statement.

(a) If $7.6 \times 237 = 1801.2$, then $0.76 \times 2370 = $ ⬚

and $1801.2 \div 76 = $ ⬚ .

(b) $3.89 \times 20 = 38.9 \times $ ⬚ $23 \div 110 = 2.3 \div $ ⬚

$2.8 \times 34 = $ ⬚ $\times 17$

$0.9 \div 15 = 9 \div $ ⬚ $= 3 \div$ ⬚ $=$ ⬚ $\div 100$

(c) If $A \times 1.5 = B \times 1.1 = C \times 0.3 = D \times 0.1$ (A, B, C and D are all greater than 0), then the greatest of these four numbers is ⬚ and the least number is ⬚ .

(d) $54 \times 3.2 + 32 + 32 \times $ ⬚ $= 320$

5 Multiple choice questions. (For each question, choose the correct answer and write the letter in the box.)

(a) When $a = 128 \times 0.25$, and $b = 128 \times 0.2$, the relationship between a and b is ⬚ .

    A. $a > b$ **B.** $a = b$ **C.** $a < b$ **D.** uncertain

(b) In all the decimal numbers with three decimal places that are rounded to 0.80, the difference between the greatest and the least is ⬚ .

    A. 0.004   **B.** 0.005   **C.** 0.009   **D.** 0.01

Challenge and extension questions

6 A lorry travelled 18 km in 0.6 hours. How many kilometres did the lorry travel in one minute? How many minutes would it take the lorry to travel 1 km?

7 Short division is a shorthand method for long division. Observe the following example carefully and complete the short division as indicated below. (For question (b), express the answer as a whole number plus a remainder.)

Example: 123.5 ÷ 5   (a) 279.6 ÷ 12   (b) 923 ÷ 7

```
        2 4 . 7
    5 ) 1 2² 3³ . 5
```

Answer: 24.7

```
    12 ) 2 7 9 . 6
```

Answer: ☐

```
    7 ) 9 2 3
```

Answer: ☐

5.2 Operations of decimals (2)

Learning objective Solve calculation problems involving decimal numbers

Basic questions

1 Work these out mentally. Write the answers.

(a) 2 − 0.55 + 0.45 = ☐

(b) 14.4 + 4.4 × 6 = ☐

(c) 0.4 + 0.6 ÷ 5 = ☐

(d) 1.8 ÷ 5 × 20 = ☐

2 Work these out step by step.

(a) 21.45 − 2.45 × 4 + 7.8

(b) (2.82 + 2.8 × 9) ÷ (0.6 + 2.4)

(c) 72 × 0.75 − 36.36 ÷ 18

(d) 5.5 ÷ 5 × 2 ÷ (2.82 + 2.18)

(e) 16.73 + 3.75 ÷ 75 × 48

(f) (4.5 − 0.45) ÷ (0.1 + 0.3 × 3)

(g) $7.85 + (3.9 - 3.51) \div 3$

(h) $(36 \times 1.25 + 3) \times 0.25 - 1.28$

3 Calculate smartly.

(a) $28.7 - 4.32 - 2.68$

(b) $82.6 - 22.6 - 41.19 + 1.19$

(c) $0.5 \times 1.25 \times 2.5 \times 64$

(d) $79 \times 1.25 + 1.25$

(e) $4.75 \div 3 + 2.3 \div 3$

(f) $4.8 + 4.8 \times 7 - 4.8 \times 6$

(g) $1.64 + 99 \times 4.5 + 2.86$

(h) $5.5 \times 7 - 2 \times 5.5 + 4.5 \times 5$

4 A factory receives an order for some work suits. According to its original plan, it needs 24 workers for 12.5 days to finish the job, and each worker will process 140 suits every day. How many suits are there in this order?

Now, under the new plan, each worker processes 10 more suits every day so the task can be completed 2.5 days earlier than the original plan. How many more workers are needed?

Challenge and extension question

5 The table shows a summary of the gas and electricity bill of Jason's family for the last quarter of 2018. Answer the questions below based on the table (excluding other charges, such as standing charges and VAT).

	Usage	Unit price
Gas (Unit: m³)	764	44.21p
Electricity (Unit: kWh)	963	12.02p

(a) How much was the monthly charge for the usage of gas?

(b) How much was the monthly charge for the usage of electricity?

(c) How much was the total charge for the family to pay for the use of gas and electricity in the quarter?

(d) Can you pose a question based on the information given in the table? Write it down and give your answer.

_____

_____

5.3 Simplification and equations (1)

Learning objective Solve algebraic problems

Basic questions

1 Simplify the following expressions.

(a) $3a + 2a =$ _____

(b) $12y \div 3 =$ _____

(c) $8b - 7b =$ _____

(d) $7x - 4x \div 2 =$ _____

(e) $2x \times 8 - x =$ _____

(f) $4a - 7b - 3a - 5b =$ _____

2 Use different methods to solve the equations.

(a) $9x \div 5 = 3.6$

Method 1:

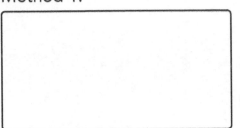

Method 2:

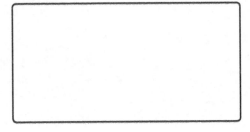

(b) $10(x + 0.15) = 5$

Method 1:

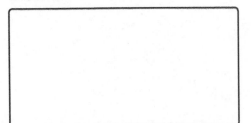

Method 2:

3 Solve the equations. (Check the answers to the questions marked with *.)

(a) $6(x - 1.5) = 15$

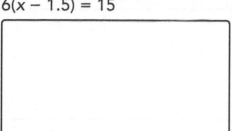

(b) $5(x + 0.24) = 2.65$

(c) $13x - 27 = 10x$

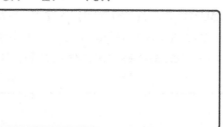

(d) $17x - 4 \times 2.7 = 57.2$

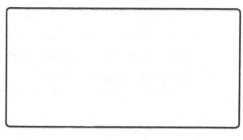

(e) $*3(2x - 5) \div 12 = 2.5$

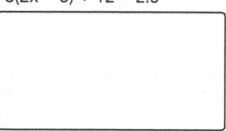

(f) $*11(x - 3) = 5.5 \times 4$

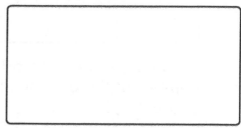

4 In each question below, write the equation and then find the solution.

(a) Three times a number is divided by 2 and the quotient is 2.4. Find this number.

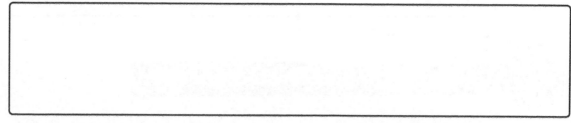

(b) Five times the sum of Number A and Number B is 19.25. Number B is 1.28. Find Number A.

(c) Number A is 1.5. Twelve times Number A equals 24 times the difference between Number B and 2.25. Find Number B.

5 Solve these problems.

(a) A family in a village cultivated 1580 m² of potato crop last year, which was 120 m² less than twice that of the area the year before last. How many square metres was the area of potatoes cultivated by the family the year before last?

(b) A school bought 22 footballs, which was 8 more than half the number of the basketballs it bought. How many basketballs did the school buy?

Challenge and extension question

6 Fill in the missing numbers.

(a) ($\boxed{}$ − 3.6) ÷ 5 × 7 + 4.8 = 7.6

(b) [12 × ($\boxed{}$ + 6) − 1.5] ÷ 4 = 102

5.4 Simplification and equations (2)

Learning objective Solve algebraic problems

Basic questions

1 Simplify and then evaluate.

(a) When $x = 3.5$, find the value of $5x + 3x \times 5$.

(b) When $m = 10$ and $n = 4.8$, find the value of $8m + 3n \div 2 - 5m$.

2 Solve the equations. (Check the answers to the questions marked with *.)

(a) $25x \div 5 + 14 = 24$

(b) $3(8 + x) \div 2 = 18$

(c) $35 - x = 4x$

(d) $5(2x + 3) = 20$

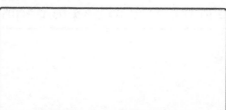

(e) (4x + 6) ÷ 3 = 7

(f) 90 − 4(x + 1) = 10

(g) *3(9x − 5x) = 27

(h) *9.6 ÷ (6x − 1.6) = 4.8

3 Use equations to solve these problems.

(a) A pigeon can fly 4.8 km in 0.4 hours. Based on this speed, how many hours does the pigeon need to fly 18 km?

(b) A shop received a batch of apples and pears. The weight of apples was 270 kg, which was 24 kg more than 1.5 times that of the pears. How many kilograms of pears did the shop receive?

(c) There are two bags of corn. The first bag weighs 150 kg. If 30 kg of corn is taken out of the first bag and put into the second bag, then the remaining corn in the first bag will be 10 kg more than the second bag. How many kilograms of corn were in the second bag to begin with?

(d) There are 6 boxes of sweets of equal mass. If 0.45 kg of sweets are taken out of each box, then the remaining sweets in the 6 boxes weigh 3.3 kg in total. How many kilograms of sweets were in each box to begin with?

Challenge and extension questions

4 If $a \blacktriangle b$ represents $(2a - b) \times b$, for example, $4 \blacktriangle 5 = (2 \times 4 - 5) \times 5 = 3 \times 5 = 15$, then, when $a \blacktriangle 3 = 24$, what is a?

5 Some Year 6 pupils participated in a mathematics competition. Among the winners, the number of boys is 3 more than 1.5 times the number of girls. The number of boys is also 2 less than twice the number of girls. How many winners are there altogether?

5.5 More equation problems

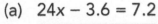 **Learning objective** Solve problems involving equations

 Basic questions

1 Solve the equations.

(a) $24x - 3.6 = 7.2$

(b) $0.4 + 5x = 2.4$

(c) $x + 5.8 - 2.5 = 9.4$

(d) $15x - 1.2 = 4.8$

2 Shanee bought 4 notebooks in a stationery shop with £50 and received £6 in change. How much did each notebook cost?

3 Mr Webb would like to buy 80 ropes in a sports shop with £200, but he is short of £40 to buy them. How much does each rope cost?

4 A gas company has 1240 employees, which is 40 more than 6 times the number of employees it had 5 years ago. How many employees did the company have 5 years ago?

5 An orchard has 370 peach trees, which is 32 fewer than 3 times the number of apricot trees it has. How many apricot trees are there in the orchard?

6 The perimeter of a rectangle is 32 cm and its length is 9.5 cm. Find the width of the rectangle.

7 Ahmed had £30 and he first bought a pack of chocolate truffles. The change he got was exactly enough to buy 2 packs of cashew nuts at £6.80 each. How much did he pay for the pack of chocolate truffles?

8 Mr Kumar bought 16 pairs of table tennis bats and 12 basketballs for £800. Each pair of table tennis bats was £27.50. How much did one basketball cost?

9 Factory A has 148 tonnes of steel and Factory B has 112 tonnes of steel. If Factory A uses 18 tonnes per day and Factory B uses 12 tonnes per day, after how many days will the two factories have the same quantity of steel?

Challenge and extension question

10 A school bought 5 tables and 8 chairs for £375. A table costs £10 more than a chair. How much does each table cost? How much does each chair cost?

5.6 Calculating the areas of shapes

Learning objective Solve problems involving the area of shapes

Basic questions

1 The figures below show a triangle and a parallelogram. Draw the height corresponding to the base indicated in each figure.

base

base

2 Calculate the area of each figure. (Unit: cm)

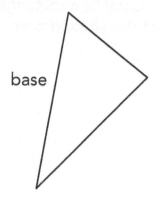

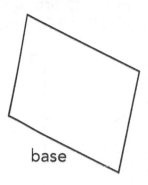

Area = ⬚   Area = ⬚

3 The area of a parallelogram is 36 cm². The base is 4.5 cm. Find the height of the parallelogram.

4 Find the area of the shaded parts in the figure. (Unit: cm)

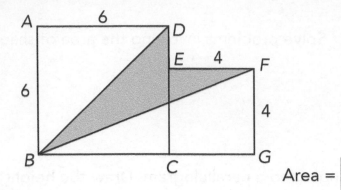

Area = ⬚

5 The perimeter of square ABCD is 24 cm. CEGF is a rectangle, EC = 2BE and DF = 2FC. Find the area of the shaded part.

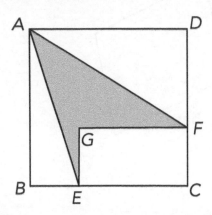

Area = ⬚

Challenge and extension question

6 The figure below shows a square ABCD with a side length 20 cm, DE = 3AE, and BF = 4AF. Find the area of triangle CEF.

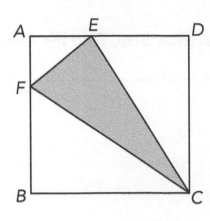

Area = ⬚

5.7 Mathematics plaza – calculation of time

 Learning objective Solve problems involving calculating time differences

 Basic questions

1 Calculate the answers.

(a) 36 minutes − 13 minutes 30 seconds = _____

(b) 5 hours 28 minutes + 3 hours 32 minutes = _____

(c) 9 minutes 14 seconds − 5 minutes 48 seconds + 1 minute 25 seconds

= _____

2 (a) Jet arrived at a Community Youth Club at 10:45 a.m. and stayed there for 48 minutes. At what time did Jet leave the club?

(b) It takes 83 minutes to travel from Southampton Central Station to London Waterloo Station by train. If the train reaches London Waterloo Station at 9:53 a.m., at what time did the train leave Southampton Central Station?

3 (a) Fatima arrives at school at five to nine in the morning. The first lesson starts at a quarter past nine. She has four lessons to attend, each lasting 45 minutes. There is a 20-minute break between the second and third lessons, and the last lesson is followed by lunch for 30 minutes. At what time does lunch break end?

(b) One day, Maria went hiking. It took her 1 hour and 45 minutes to walk up to the top of a hill. After a half-hour break she walked back, which took 1 hour and 26 minutes. How much time did the whole hike take?

4 It takes 45 minutes to bake a cake. If a cake has been baking for 6 minutes and 32 seconds, how much time is still needed?

5 Kim has a good habit of reading from half past six to twenty past eight every evening. Accordingly, how much time does she spend on reading in the evenings in a week (7 days) in total?

6 Jack is going to a concert, which starts at 19:30 and ends at 21:30. It takes 45 minutes to get there and 40 minutes to get back home. At what time should he leave home for the concert if he needs to arrive 10 minutes before it starts? At what time will he arrive at home if he goes home immediately after the concert?

Challenge and extension question

7 A film lasts 95 minutes. A cinema plans to start the film at 9 o'clock in the morning. The interval between two shows is 20 minutes for the films shown before 16:00 and 15 minutes for those after 16:00. Based on this plan, answer the following questions.

(a) The cinema closes at 22:00. Work out at what time the last film should end.

(b) The ticket price of a show varies depending on the show time. Each ticket is £8 for shows before 16:00 and £10 for those after 16:00. The cinema has a total of 600 seats and, on average, there are 100 seats unsold for each show. If a film is on show for one week (7 days), how much money does the cinema earn by selling the tickets?

Chapter 5 test

1 Work out the answers mentally. Write the answers.

(a) $1 \div (0.25 \times 8) = \boxed{}$

(b) $8.7 \times 0.5 \times 20 = \boxed{}$

(c) $7.6 - 1.5 \times 4 = \boxed{}$

(d) $(5.4 + 7.1) \div 2 = \boxed{}$

2 Use the column method to calculate the following.

(a) 7.24×58

(b) $62.26 \div 11$

3 Solve the equations.

(a) $7.5x - 4.9 = 2.6$

(b) $16(x + 4.5) = 4.8 \times 20$

4 Work these out step by step. (Calculate smartly if possible.)

(a) $5.8 - 0.8 \div (1.5 + 0.5)$

(b) $26.22 - 58.4 - 41.6 + 73.78$

(c) $1.25 \times 16 \times 0.25$

(d) $(3.6 \times 4 + 8 \times 3.6) \div 6$

(e) $2.25 \div (75 \div 25 + 1.5)$

(f) $5.6 \times [9.91 + (10.04 - 9.59) \div 5]$

5 A school canteen received a delivery of cooking oil. If it uses 30 kg of the cooking oil every day, the cooking oil could last 40 days. With fewer customers, the canteen actually saves 5 kg of the cooking oil every day. Therefore, how many days can the cooking oil last?

6 Jason has £30. After he bought 4 notebooks, he spent the rest of the money on 2 packs of pens at £7 each. How much did one notebook cost?

7 A company planned to buy 2 desks at £480 each. Later, it decided to use the money to buy 7 chairs instead of desks. However, the money was not enough and the company had to pay £90 more. How much did one chair cost?

8 A school bought 16 sets of badminton racquets, which cost £800 in total. The price of each set of badminton racquets is £2 more than 8 times the price of a skipping rope. What is the price of a skipping rope?

9 The starting fare of a taxi in a foreign city is a flat £10 for a maximum of 3 km. When it travels more than 3 km but less than 10 km, each additional km (or less) travelled is charged £2. When it travels over 10 km, each additional km (or less) is charged £3.

(a) Jamal took a taxi and travelled 8 km, how much did he have to pay?

(b) Serena took a taxi and the fare shown was £30. How many kilometres did she travel in the taxi?

10 Find the area of the figure. (Unit: m)

3.5

10

8

Area =

11 The figure shows three identical small squares and two identical large squares forming a rectangle *ABCD*. Fill in the spaces below.

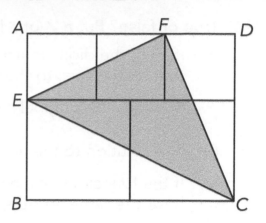

(a) If the side length of the small square is

a cm, then its area is _____ cm².

The side length of the large square

is _____ cm².

(b) If the perimeter of rectangle *ABCD* is 44 cm, then its area

is _____ cm². The area of triangle *ECF* is _____ cm².

Questions 12 to 15 are multiple choice questions. (For each question, choose the correct answer and write the letter in the box.)

12 Round 0.987 65 to the nearest thousandth. It is ☐.

   A. 0.987   **B.** 0.988   **C.** 0.998   **D.** 1.000

13 If $a \div b > a \, (a > 0)$, then *b* could be ☐.

   A. 0   **B.** 0.3   **C.** 1   **D.** 2.5

14 The areas of a triangle and a parallelogram are the same. The base of the parallelogram is twice the base of the triangle. If the height of the parallelogram is 4 cm, then the height of the triangle is ☐.

   A. 2 cm   **B.** 4 cm   **C.** 8 cm   **D.** 16 cm

15 In the following four statements, the incorrect one is ☐.

   A. A rectangle is also a parallelogram.

   B. A rectangle is a symmetrical figure.

   C. A triangle can have three acute angles.

   D. A right-angled triangle can have two right angles.

16 True or false? Put a ✓ or ✗ in each box.

(a) If all of the digits in a number are moved one place to the right and then two places to the left across the decimal point, the new number is 10 times the original number. ☐

(b) The solution to the equation $18x + 2.4 = 3.6 - 6x$ is $x = 0.5$. ☐

(c) If the four sides of a parallelogram and the four sides of a rectangle have the same lengths, then the parallelogram and the rectangle must have the same area too. ☐

(d) Tom watched TV from 19:55 to 21:05. He spent 1 hour 50 minutes watching TV. ☐

For questions 17 to 20, fill in the missing numbers.

17 If $64.8 ÷ 18 = 3.6$, then $18 × 0.36 =$ ☐ , $64.8 ÷ 360 =$ ☐ .

18 If $5.4 × 38 + 3.8 × ▇ = 380$, then the number in the ▇ is ☐

19 In the figure on the right, both triangle *ABC* and triangle *CDE* are right-angled isosceles triangles. If $DE = 12\,cm$, then the area of

triangle *CDE* is ☐ cm^2, the area of the

shaded square is ☐ cm^2, and the area

of triangle *ABC* is ☐ cm^2.

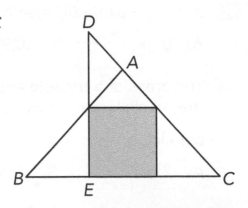

20 A rope has half its length cut off. It then has *m* metres cut off. Then there are 2 metres left. Using an expression to represent the total length of the rope, including the letter *m*, it was ☐ metres.

When $m = 5.5$ metres, the total length of the rope was ☐ metres.